# Strongly Interacting Quantum Systems, Volume 2

Many-body physics

Online at: https://doi.org/10.1088/978-0-7503-3091-6

# IOP Series in Quantum Technology

**Series Editor**: **Barry Garraway** (School of Mathematical and Physical Sciences, University of Sussex, UK), **Barry Sanders** (Institute for Quantum Science and Technology, University of Calgary, Canada) and **Lincoln Carr** (Quantum Engineering Program, Colorado School of Mines, USA)

## About the Series

The IOP Series in Quantum Technology is dedicated to bringing together the most up to date texts and reference books from across the emerging field of quantum science and its technological applications. Prepared by leading experts, the series is intended for graduate students and researchers either already working in or intending to enter the field. The series seeks (but is not restricted to) publications in the following topics:

- Quantum biology
- Quantum communication
- Quantum computation
- Quantum control
- Quantum cryptography
- Quantum engineering
- Quantum machine learning and intelligence
- Quantum materials
- Quantum metrology
- Quantum optics
- Quantum sensing
- Quantum simulation
- Quantum software, algorithms and code
- Quantum thermodynamics
- Hybrid quantum systems

A full list of titles published in this series can be found here: https://iopscience.iop.org/bookListInfo/iop-series-in-quantum-technology.

# Strongly Interacting Quantum Systems, Volume 2

## Many-body physics

**Manuel Valiente**
*Universidad de Murcia, Murcia, Spain*

**Nikolaj Thomas Zinner**
*Aarhus University, Aarhus, Denmark*

**IOP** Publishing, Bristol, UK

ISBN    978-0-7503-3091-6 (ebook)
ISBN    978-0-7503-3089-3 (print)
ISBN    978-0-7503-3092-3 (myPrint)
ISBN    978-0-7503-3090-9 (mobi)

DOI    10.1088/978-0-7503-3091-6

Version: 20251001

IOP ebooks

British Library Cataloguing-in-Publication Data: A catalogue record for this book is available from the British Library.

Published by IOP Publishing, wholly owned by The Institute of Physics, London

IOP Publishing, No.2 The Distillery, Glassfields, Avon Street, Bristol, BS2 0GR, UK

US Office: IOP Publishing, Inc., 190 North Independence Mall West, Suite 601, Philadelphia, PA 19106, USA

*A mis niños: Álvaro, Antonio, David H y Manuel K.*

—Manuel Valiente

*Til Tine, Krümmel og Chloé*

—Nikolaj Thomas Zinner

# Contents

# Preface

This book is the second volume of a series of books on the topic of strongly interacting quantum systems. The first volume appeared in 2023, and deals with few-body quantum systems. Throughout this book, we shall refer to it simply as volume 1.

This second volume is concerned with strongly interacting quantum many-body systems. There are a vast amount of basic and advanced textbooks available on quantum many-body physics. Almost any preface deals with this issue, and answers it with the question 'another book on many-body physics?'. The answer, always affirmative, goes on to justify the need for yet another book on the topic. We shall do no such thing. This is not a book on many-body physics, but on the use of few-body machinery to tackle non-perturbative physics in many-body systems. It often turns out that few-particle solutions are sufficient to explain a plethora of many-body phenomenology. Sometimes, quantitative predictions can be made, while other times relations or qualitative—but highly non-perturbative—estimates are made. This field is, admittedly, quite niche, and we do not know of any other textbook on the subject. It can be difficult at times, and so we have included as many details in our derivations and calculations as possible, following the same philosophy as in volume 1. Although not always necessary, we recommend the reader to carefully study volume 1 before attempting to read this book.

We have chosen a number of topics that we think are most representative of the approach we follow. We have left out important topics and systems that are very interesting, but we are sure that these pages will help the reader to develop this field further, should they be interested in doing so.

We thank our Editor at IOP Publishing, Phoebe Hooper, for her help and persistence. We could neither have started nor finished this project without the support of many colleagues with whom we have discussed and collaborated in this and other topics, and the funding agencies that have supported our work. We would like to thank our colleagues C W Duncan, P Öhberg, L G Phillips, H Zhai, V Pastukhov, A G Volosniev, O V Marchukov, D V Fedorov, A S Jensen, J R Armstrong, H-W Hammer, K Mølmer, D Petrosyan, A Saenz, K Fraser, B Juliá-Díaz, I Morera, M A García-March, S Mukherjee, E Andersson, M J Edmonds, L Santos, Y Nishida, Y Sekino, F F Bellotti, T Frederico, M T Yamashita, L Tomio, J B Hofmann, X Chen, N Barnea, B Bazak, D Pérez-Cruz and X Cui.

We acknowledge support from the Ramón y Cajal programme, Spanish Ministry of Science and Innovation through grant number RYC2020-029961-I, as well as the national research and development grant PID2021-126039NA-I00.

**Manuel Valiente**
Murcia, Spain

**Nikolaj Thomas Zinner**
Aarhus, Denmark

# Author biographies

## Manuel Valiente

**Manuel Valiente** is currently Associate Professor of Physics at the University of Murcia, Spain. Previously he held a Ramón y Cajal Fellowship at the Universities of Murcia and La Laguna, Spain. He was a postdoctoral researcher at several institutions around the world, and previously held a Faculty position at Tsinghua University in Beijing. He obtained a PhD in Physics at Humboldt-Universität zu Berlin in Germany in 2010.

## Nikolaj Thomas Zinner

**Nikolaj Thomas Zinner** is Professor of theoretical physics at the Department of Physics at Aarhus University, Denmark, where he works mainly in quantum dynamics and quantum technology. He was a postdoctoral fellow at Harvard University and at the Niels Bohr Institute before becoming a faculty member at Aarhus University. Nikolaj obtained his PhD in theoretical nuclear physics in 2007 from Aarhus University.

**IOP** Publishing

# Strongly Interacting Quantum Systems, Volume 2
## Many-body physics
**Manuel Valiente and Nikolaj Thomas Zinner**

# Chapter 1

# Short-distance universality

One of the most important and useful concepts in physics is the separation of scales. Within a complex system, several time, distance or energy scales may be present, and very distinct phenomena may be associated with the different physical scales in the problem. As an example from everyday life, we may stare at a light bulb. Our brains will interpret the light intensity that reaches our eyes essentially as a constant. However, this is not the whole reality of the situation. The AC current flowing through it oscillates at 50 Hz here in Europe, which lies around the critical flicker fusion frequency for humans. You would certainly be very uncomfortable studying this book with 20 Hz current! This very simple example tells us that, depending on the scale we are probing, we may see different levels of detail or perceive different physical phenomena. In this chapter, we will explore the separation of distance or momentum scales in quantum many-body systems and, moreover, extract important information that relates, non-perturbatively, global properties of these systems to their short-distance, or large-momentum behaviours.

We will begin with a number of general properties of two-body systems, which were left out of Volume 1 as they are most relevant here. Then, we will move on to set a number of important length scales in atomic and nuclear physics. Finally, we shall derive many useful, non-perturbative relations that are commonly englobed in the topic of short-distance universality, both in their weak and strong versions.

## 1.1 Two-body physics at short distances

Let us consider two identical, non-relativistic particles with mass $m$ interacting via a pairwise potential $V(\mathbf{r})$, with $\mathbf{r} = \mathbf{r}_1 - \mathbf{r}_2$ their relative position. In $D$ spatial dimensions, the stationary Schrödinger equation in the relative coordinate is given by

$$-\frac{\hbar^2}{2\mu}\,\nabla^2\,\psi + V(\mathbf{r})\psi = E\psi, \tag{1.1}$$

doi:10.1088/978-0-7503-3091-6ch1

where $\mu = m/2$ is the reduced mass, and $\nabla^2$ is the $D$-dimensional Laplacian. If the interaction is hyperspherically symmetric, as we will assume from here on, then the relative wave functions separate as

$$\psi(\mathbf{r}) = R(r)\, Y_{\vec{l}}(\vec{\theta}). \tag{1.2}$$

Above, $r$ is the hyperradius, $\vec{l}$ and $\vec{\theta}$ represent, respectively, $D - 1$ angular momentum quantum numbers, and $D - 1$ angles, while $Y_{\vec{l}}$ are hyperspherical harmonics. For $D = 3$, we have $Y_{\vec{l}} \equiv Y_{l,m}(\theta, \phi)$; for $D = 2$, $Y_{\vec{l}} \equiv \exp(im\phi)/\sqrt{2\pi}$; in one spatial dimension, angular momentum channels reduce to parity, and we have $Y_{\vec{l}} = (\mathrm{sgn}(x))^{\mathrm{P}}$, where $\mathrm{P} = 0, 1$ is the parity quantum number.

The separation of parity and radial coordinate for $D = 1$ is typically uncommon, so we derive it here, and it will prove useful in the following chapter. The radial coordinate is simply $r = |x|$. The Laplacian, since there is only one degree of freedom, is invariant with respect to the change $x = \mathrm{sgn}(x)|x|$, except at the origin, at which an appropriate boundary condition must be set. We have

$$\nabla^2 = \frac{\partial^2}{\partial x^2} = \frac{\partial^2}{\partial |x|^2}, \quad \forall\, x \neq 0. \tag{1.3}$$

Therefore, if the two-body interaction is regular, in the even channel, the Schrödinger equation is reduced to

$$-\frac{\hbar^2}{2\mu} R''(r) + V(r)R(r) = ER(r), \quad r > 0, \tag{1.4}$$

together with the boundary conditions $R(0) = \varphi_0$ for some finite $\varphi_0 \in \mathbb{C}$ and either $\int_0^\infty dr R(r) < \infty$ (bound states) or $|R(r)|$ bounded as $r \to \infty$ (scattering states). In the odd channel, care must be taken *a priori*, since the behaviour at $x = 0$ is relevant. If the interaction potential $V(r)$ is regular, continuity of the wave function at $x = 0$ together with odd parity implies $R(0) = 0$ and, therefore, the Schrödinger equation becomes equation (1.4) with identical boundary conditions as in the even-wave case, but with $\varphi_0 = 0$. If the interaction is not regular, then we write the wave function as $\psi(x) = \mathrm{sgn}(x)R(r)$, and after application of the Laplacian we see that

$$\frac{\partial^2}{\partial x^2}[\mathrm{sgn}(x)R(r)] = 2\delta(r)R'(r) + \mathrm{sgn}(x)R''(r), \tag{1.5}$$

where $\delta(r)$ is the Dirac delta function, and where we have used the distributional identity $\delta'(r) = -\delta(r)\partial/\partial r$. The 'radial' Schrödinger equation reads, in this case $(r \geqslant 0)$

$$-\frac{\hbar^2}{2\mu} R''(r) - \frac{\hbar^2}{\mu}\delta(r)R'(r) + V(r)R(r) = ER(r). \tag{1.6}$$

Above, we have implicitly assumed that $\mathrm{sgn}(0) = 1$, a choice that we term Shirokov's choice, for reasons that will be clear in the following chapter.

Formally, equation (1.6) can have a solution if we split the interaction potential as $V(r) = V_{\text{reg}}(r) + V_{\text{I}}(r)$, with $V_{\text{I}}(r) = (\hbar^2/\mu)\delta(r)\partial/\partial_r$. In this case, it reduces to equation (1.4) with $V(r)$ replaced with $V_{\text{reg}}(r)$, and with the boundary condition $R(0) = \varphi_0$. That is, the problem with an irregular potential of the form described above leads to an even-wave (or bosonic) problem with identical properties. The main consequence is that the full odd-wave function $\psi(x) = \text{sgn}(x)R(|x|)$ is discontinuous at $x = 0$.

In two dimensions, the radial wave function obeys the following Schrödinger equation for angular quantum number $m$,

$$-\frac{\hbar^2}{2\mu}\left[ R''(r) + \frac{1}{r}R'(r) - \frac{m^2}{r^2}R(r) \right] + V(r)R(r) = ER(r). \tag{1.7}$$

In three dimensions, rewriting, as usual, the radial wave function $R(r) = u(r)/r$ for $r > 0$ and $R(0) = \lim\limits_{r \to 0^+} u(r)/r$, the Schrödinger equation with angular momentum quantum numbers $(l, m)$ takes the form

$$-\frac{\hbar^2}{2\mu}u''(r) + \left[ V(r) + \frac{\hbar^2}{2\mu}\frac{l(l+1)}{r^2} \right]u(r) = Eu(r), \tag{1.8}$$

with the boundary condition $u(0) = 0$. Note that this is in one-to-one correspondence with the regular, odd-wave, one-dimensional case.

We take a deeper look at the three-dimensional problem, and in the $s$-wave channel ($l = 0$). Firstly, we notice that at low energies and short distances ($r \leqslant r_0$), with

$$\left| \frac{V(r) - E}{\hbar^2/2\mu r_0^2} \right| \ll 1, \tag{1.9}$$

we may neglect the energy in equation (1.8) and work with $E \equiv 0$. Then, if $u(r) \sim \bar{u}(r)$ as $r \to 0$ is the asymptotic short-distance wave function, we have

$$\frac{\bar{u}''(r)}{\bar{u}(r)} \approx \frac{2\mu}{\hbar^2}V(r), \quad r \to 0. \tag{1.10}$$

If the interaction potential is smooth at short distances (which is usually unphysical), we can extract the short-distance asymptotics up to an arbitrary constant. For instance, let us consider an exponential interaction of the form

$$V(r) = \frac{\hbar^2}{2\mu}v_0 e^{-\lambda r}. \tag{1.11}$$

Expanding $V(r)$ in Taylor series, $2\mu V(r)/\hbar^2 = v_0 \sum_{k \geqslant 0} \lambda^k r^k/k!$, as well as $\bar{u}(r) = \sum_{m \geqslant 1} u_m r^m$, we get, up to the fourth power,

$$\bar{u}(r) = u_1\left[ r + \frac{v_0}{6}r^3 - \frac{v_0\lambda}{12}r^4 \right] + O(r^5). \tag{1.12}$$

To this order, the only finite-energy correction occurs at $O(r^3)$ where, instead of the coefficient above, we obtain $u_1(v_0 - \varepsilon)/6$, with $\varepsilon = 2\mu E/\hbar^2$. Therefore, neglecting the energy is a valid assumption for $|v_0 - \varepsilon|r_0^2 \ll 1$, as expected.

All of the above analysis is neat and correct. However, we need to ask ourselves whether the distances at which an expansion such as that in equation (1.12) are valid also correspond to distances that can be probed and, moreover, whether that would be at all useful. While we anticipate that such expansions will not be very physically meaningful, we will postpone the analysis until after the following and important theoretical interlude.

We continue setting $l = 0$ ($s$-waves) and zero energy. Now, however, we are going to assume that the distance $r$ between the particles is short as compared with some other length scale, $R_*$, to be specified later. We further assume that the particles are not so close together that the microscopic details of the interaction potential play a role. That is, the particles lie at distances $r \gg r_0$, with $r_0$ defined in equation (1.9), and $r < R_*$, whatever $R_*$ might be. At zero energy, and if $V(r)$ has a strict finite range ($V(r) = 0$ for $r > R_0$), then the wave function is in the long-distance asymptotic regime, that is, $u(r) \propto r - a$, with $a$ the $s$-wave scattering length. It seems, then, reasonable, that the short- and long-range behaviours of the *two-body* wave function may overlap depending on the length scales at which the system is probed. Realistic two-body interactions, however, do not have a strictly finite range, but have long-distance tails. In neutral atomic collisions, for instance, van der Waals tails, which are proportional to $r^{-6}$, are present. In general, let us consider asymptotic potentials, $V_\infty(r)$, such that, if $r \geqslant r_0$ for some distance $r_0$, they satisfy

$$\left| \frac{V(r) - V_\infty(r)}{\hbar^2/2\mu r^2} \right| \ll 1. \tag{1.13}$$

Considering that the above asymptotic condition is fulfilled and that $r < R_*$, the zero-energy Schrödinger equation is approximated as

$$-\frac{\hbar^2}{2\mu}\bar{u}''(r) + V_\infty(r)\bar{u}(r) = 0, \quad r_0 \leqslant r < R_*. \tag{1.14}$$

Above, and since we are only considering two particles, we may assume simply that $r \geqslant r_0$. We consider power-law tails, that is,

$$V_\infty(r) = \frac{\hbar^2 g_\nu}{2\mu}r^{-\nu}, \tag{1.15}$$

with $\nu > 0$. We will see shortly further conditions on $\nu$. Instead of the fully asymptotic wave function, we consider the following ansatz

$$\bar{u}(r) = r - a(r), \tag{1.16}$$

where $\lim_{r \to \infty} a(r) = a$. Because the tails are power laws, we may expand $a(r)$ as

$$a(r) = a + \frac{a_1}{r^\eta} + \cdots, \tag{1.17}$$

with $\eta > 0$. Inserting the asymptotic wave function into the asymptotic Schrödinger equation, and keeping order-by-order consistency, we find that $\eta = \nu - 3$, which implies that $\nu > 3$, and

$$a(r) = a + \frac{g_\nu}{(\nu - 3)(\nu - 2)} \frac{1}{r^{\nu-3}} + \cdots, \quad \nu > 3. \tag{1.18}$$

The above result coincides with the fact that no finite $s$-wave scattering length is found for interactions that, at long distances, feature power laws that decay slower than $r^{-3}$. This result, however, can be strengthened when the tails are not monotonic. For instance, the Ruderman–Kittel–Kasuya–Yosida (RKKY) interaction, which is very important for the phenomenon of giant magnetoresistance, features a power-law tail (with $\nu = 3$) modulated by a sinusoidal function of the interparticle distance. However, two non-relativistic particles interacting via such potential do feature a finite scattering length. To see this, consider a spherically symmetric asymptotic interaction of the form

$$V_\infty(r) = \frac{\hbar^2 g_\nu}{2\mu} \cos(qr) r^{-\nu}, \tag{1.19}$$

where $q$ is a wave number with dimensions of inverse length. We use now the following ansatz

$$a(r) = a_1 \cos(qr) r^{-\eta}. \tag{1.20}$$

Proceeding just as for the monotonic power-law tails, we obtain $\eta = \nu - 1$, implying $\nu > 1$, and

$$a(r) = a - \frac{g_\nu}{q^2} \frac{\cos(qr)}{r^{\nu-1}}, \quad \nu > 1. \tag{1.21}$$

That is, the sinusoidal modulation leads to a more relaxed condition on the tails for the scattering length to remain well defined.

We can also repeat the asymptotic analysis for two-body bound states. In this case, if the binding energy is $B_2 = \hbar^2 \lambda^2 / 2\mu$, with $\lambda > 0$, then we use the ansatz $\bar{u}(r) = \exp(-\lambda(r)r)$, with $\lim_{r\to\infty} \lambda(r) = \lambda$. For power-law asymptotic potentials, we expand $\lambda(r)$ as

$$\lambda(r) = \lambda + \frac{\lambda_1}{r^\nu} + \cdots, \tag{1.22}$$

with $\nu > 0$. Proceeding analogously to zero-energy scattering, we find

$$\lambda(r) = \lambda + \frac{g_\nu}{2(1 - \nu)\lambda} \frac{1}{r^\nu} + \cdots, \quad \nu > 1. \tag{1.23}$$

The critical value, $\nu = 1$, corresponds to the Coulomb interaction, for which $\lambda(r) = \lambda$, as is well-known. We have

$$\bar{u}(r) \propto \exp\left[-\lambda r + \frac{g_6}{10\lambda r^5}\right]. \tag{1.24}$$

We can test the weak energy dependence at short ranges for the $^4He_2$ molecule. A recent experiment, reference [1], measured the radial probability density, in the relative coordinate, that is, $|u(r)|^2$, using the technique known as Coulomb explosion. We compare the experimental result at short distances with a numerical calculation at zero energy, using the Aziz interaction potential [2]. It takes the form $V(r) = \varepsilon v(x)$, with $x \equiv r/r_m$, and

$$v(x) = Ae^{-\alpha x} - \left[ \frac{C_6}{x^6} + \frac{C_8}{x^8} + \frac{C_{10}}{x^{10}} \right] F(x), \tag{1.25}$$

$$F(x) = \exp\left[-(1 - D/x)^2\right], \quad x < D, \\ = 1, \quad x \geqslant D. \tag{1.26}$$

The numerical constants for the potential are $\varepsilon/k_B = 10.8K$, where $k_B$ is Boltzmann's constant, $r_m = 2.9673$ Å, $D = 1.241\,314$, $A = 0.544\,850$, $\alpha = 13.353\,384$, $C_6 = 1.373\,2412$, $C_8 = 0.425\,3785$ and $C_{10} = 0.178\,100$. For the calculation, we used $\hbar^2/2\mu = 12.12K$ Å$^2/k_B$. In figure 1.1, we plot the experimental data for the dimer together with $|u(R)|^2$ at zero energy calculated for the Aziz potential. Clearly, they agree at short distances, and deviations beyond $R \sim 8$ Å and up until $R \sim 20$ Å can be attributed to the inadequacy of the Aziz potential at very low energies, since it overestimates the scattering length by about 30%, and the interaction is already in the asymptotic limit. To compare with the asymptotic wave function, yet at short distances, given the scattering length $a$, and the coefficients $C_6$ and $C_8$, we can proceed as before and find, to this order (see problem 1) $a(r) = a + a_1 r^{-3} + a_2 r^{-4} + a_3 r^{-5}$, obtaining

$$a_1 = -\frac{g_6}{12}, \tag{1.27}$$

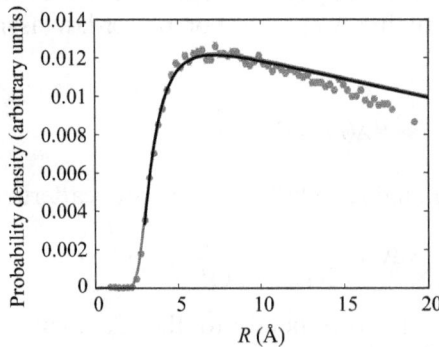

**Figure 1.1.** Radial probability density $|u(R)|^2$ for the $^4He$ dimer at short distances. Red dots are the experimental results of reference [1]; blue solid line is the zero-energy calculation with the Aziz potential, reference [2]; black solid line (from $r = 3$ Å) is the asymptotic result for the long-range part of the Aziz potential, with scattering length $a = 124.486$ Å.

$$a_2 = \frac{g_6 a}{20}, \tag{1.28}$$

$$a_3 = -\frac{g_8}{30}. \tag{1.29}$$

In figure 1.1, we see that the asymptotic wave function is accurate down to about $r_m \approx 3$ Å.

For completeness, we perform the asymptotic analysis for finite (positive) energies, for which the requirements on the tails are also relaxed. For $kr \to \infty$, with $k$ the momentum satisfying $E = \hbar^2 k^2 / 2\mu$, we may employ the following ansatz for the asymptotic wave functions

$$\bar{u}(r) = \sin(kr + \theta_k(r)), \tag{1.30}$$

where $\lim_{r \to \infty} \theta_k(r) = \theta_k$ is the $s$-wave scattering phase shift at momentum $k$. Introducing equation (1.30) into the asymptotic Schrödinger equation, we obtain the following non-linear differential equation,

$$\frac{1}{2}(\theta'(r))^2 + k\theta'(r) + \frac{2\mu V_\infty(r)}{\hbar^2} = 0. \tag{1.31}$$

With the reasonable assumptions that $\lim_{r \to \infty} |\theta''(r)/V_\infty(r)| = \lim_{r \to \infty} |\theta''(r)/\theta'(r)| = 0$, equation (1.31) is solved directly as

$$\theta_k(r) = \theta_k - \frac{\mu}{\hbar^2 k} \int dr \, V_\infty(r). \tag{1.32}$$

As a convention, we choose the integration constant such that $\lim_{r \to \infty} \theta_k(r) = \theta_k$, the actual $s$-wave scattering phase shift. Clearly, equation (1.32) gives, for power-law tails $V_\infty(r)$ in equation (1.15),

$$\theta_k(r) = \theta_k + \frac{\mu}{\hbar^2 k} \frac{g_\nu}{\nu - 1} \frac{1}{r^{\nu - 1}} + \cdots, \quad \nu > 1, \tag{1.33}$$

$$\theta_k(r) = \theta_k - \frac{\mu}{\hbar^2 k} g_1 \log|2kr| + \cdots, \quad \nu = 1. \tag{1.34}$$

Sinusoidal modulations of the tails improve further the asymptotic behaviour, and we leave this as an exercise for the reader (see problem 2).

With all of the above results, we will be in a good position to study short-range universality in many-body systems. Before that, we look into important physical scales in atomic and nuclear physics.

## 1.2 A matter of scales: atoms and nuclei

What defines asymptotic regions—be it short- or long-range—depends on the particular system of interest, as well as the methods and energy ranges we use in our probes. Here, we take a few representative examples in ultracold atomic gases, as well as nuclei.

Our first example is the neutral atomic $^{87}$Rb. This is a bosonic atom, an ensemble of which was the first Bose–Einstein condensate to be produced with ultracold alkali atoms. There are several length scales that are relevant. Firstly, the mean square radius (rms) in its electronic ground state, which is nothing but rms $\equiv \langle r^2 \rangle^{1/2}$. For the Rb atoms to be considered as point-like bosons, the mean interparticle distances must be far greater than the rms. The next important length scale to consider is the van der Waals length $R_{\text{vdW}}$, which is characterised by the coefficient $C_6$ that controls the strength of the $r^{-6}$ tail of the Rb–Rb Born–Oppenheimer interaction, that is,

$$V_\infty(r) = -\frac{C_6}{r^6}. \tag{1.35}$$

The van der Waals length is defined via

$$R_{\text{vdW}} = \frac{1}{2}\left(\frac{2\mu C_6}{\hbar^2}\right)^{1/4}. \tag{1.36}$$

For $R > R_{\text{vdW}}$, the scattering wave function reaches its asymptotic form. The last scale we need, which we had denoted $R_*$, is nothing but the typical interparticle distance. This is not specific to the Rb atoms, but to the prepared density of the atomic gas. If the density is called $\rho$, then $R_*$ scales as $R_* \sim \rho^{-1/3}$.

Our second example is atomic $^{40}K$. This one is a fermion. The two spin states are accomplished by mapping $|\uparrow\rangle = |9/2, -7/2\rangle$ and $|\downarrow\rangle = |9/2, -9/2\rangle$, where the first quantum number is the total atomic spin $f$, while the second quantum number $m_f$ is the projection along the magnetic field axis. Here, the atom number density is given by $\rho = 4\pi k_F^3/3$, so that the interparticle distance $R_* \sim \rho^{-1/3}$, just as before. The van der Waals length is different from the one for $^{87}$Rb, since both mass and $C_6$ are different.

Our third example concerns large, bound clusters of $^4$He, where the interparticle distance is related to the equilibrium density $\rho_{\text{eq}}$ of the Helium droplet as $R_* \sim \rho_{\text{eq}}^{-1/3}$.

The last example is generic to atomic nuclei, which are bound clusters of neutrons and protons, with average interparticle distances $R_*$ of the order of the range of the pion-exchange interaction $r_\pi = \hbar/m_\pi c \approx 1.4$ fm.

In table 1.1, we give estimates of these quantities for all the four examples just mentioned.

**Table 1.1.** Characteristic length scales for different systems.

|  | $^{87}$Rb | $^{40}$K | $^4$He | Nuclei |
|---|---|---|---|---|
| $R_*$ | 100 nm | $3 \times 10^5$ nm | 0.5 nm | 2.4 fm |
| $R_{\text{vdW}}\|r_\pi$ | 4.4 nm | 3.4 nm | 0.26 nm | 1.4 fm |
| rms | 0.24 nm | 0.22 nm | 0.03 nm | $1.2A^{1/3}$ fm |

## 1.3 Weak short-range universality

In this section we present the weak version of short-range universality. This occurs when the length scale associated with the two-body interaction—be it the van der Waals length or the pion-exchange range—is comparable with the interparticle spacing in the system $R_*$. In the previous section, table 1.1, this is the situation in low-temperature $^4$He as well as in atomic nuclei. We will begin with the many-boson problem, corresponding to the example of $^4$He, and then study the particularities of spin-1/2 fermionic systems.

### 1.3.1 Many-boson problem

The many-body Hamiltonian for $N$ non-relativistic spinless bosons of mass $m$ interacting via pairwise potentials is given by

$$H = -\frac{\hbar^2}{2m}\sum_{i=1}^{N}\nabla_i^2 + \sum_{i<j=1}^{N}V(\mathbf{r}_i - \mathbf{r}_j). \tag{1.37}$$

If we pick any two particles aside, say particles 1 and 2, we may split the Hamiltonian as $H = H_{1,2} + H_{3,...,N} + \mathcal{H}$, where $H_{1,2}$ is the part of the Hamiltonian only involving particles 1 and 2, $H_{3,...,N}$ the part only involving particles 3 to $N$, and $\mathcal{H}$ is their coupling, that is

$$\mathcal{H} = \sum_{j=3}^{N}[V(\mathbf{r}_1 - \mathbf{r}_j) + V(\mathbf{r}_2 - \mathbf{r}_j)] \tag{1.38}$$

If the distance $|\mathbf{r}_1 - \mathbf{r}_2| \to 0$, that is, when the distance between all particles is far greater than the distance between particles 1 and 2, this separation implies that the eigenfunctions $\psi$ of the Hamiltonian factorize as

$$\psi(\mathbf{r}_1, \mathbf{r}_2,...,\mathbf{r}_N) \to \phi_2(\mathbf{r}_{12})A_2^{(N)}(\mathbf{R}_{12}; \{\mathbf{r}_k\}_{k=1}^{N}). \tag{1.39}$$

Above, $\mathbf{r}_{12} = \mathbf{r}_1 - \mathbf{r}_2$ is the relative coordinate, $\mathbf{R}_{12} = (\mathbf{r}_1 + \mathbf{r}_2)/2$ is the centre of mass coordinate for two particles, $\phi_2(\mathbf{r}_{12})$ is a two-body eigenstate, and $A_2^{(N)}$ depends on the centre of mass coordinate of the 1-2 system as well as on the rest of the coordinates of the particles. Since we have argued that, at least at low energies, the energy at which we evaluate the two-body, short-distance state, is irrelevant, we may take $\phi_2$ as the zero-energy $s$-wave state or, in the case of bound systems, such as $^4$He, the two-body $s$-wave ground state. Note that, in the case of $^{87}$Rb, the relevant many-body states for a Bose–Einstein condensate are *not* the ground state of the system: $^{87}$Rb admits large numbers of deeply bound molecular states which, at very short ranges exhibit a wildly oscillating wave functions with a large number of nodes. For bosonic helium molecules, however, the ground state is very weakly bound—with a binding energy of about $B_2 \approx 1.3$ mK—and, obviously, has no nodes. From now on, we will assume that $\phi_2$ is the two-body ground state of the system in the relative coordinate.

To continue, we define the norm of $A_2^{(N)}$ as

$$\langle A_2^{(N)} | A_2^{(N)} \rangle \equiv \int \prod_{k=3}^{N} d\mathbf{r}_k d\mathbf{R}_{12} \left| A_2^{(N)}(\mathbf{R}_{12}; \{\mathbf{r}_k\}_{k=3}^{N}) \right|^2, \tag{1.40}$$

and define the two-body contact as

$$C_2^{(N)} \equiv \frac{N(N-1)}{2} \langle A_2^{(N)} | A_2^{(N)} \rangle. \tag{1.41}$$

The above quantity is central to the study of short-range universality in quantum many-body physics. Our objective now is to relate a number of measurable quantities at short distances with the aforementioned contact.

**Pair density function.** If the pairwise interaction potential is spherically symmetric, it is convenient to define the pair density operator $\hat{\rho}_2^{(N)}(r)$ as

$$\hat{\rho}_2^{(N)}(r) = \frac{1}{r^2} \sum_{i<j=1}^{N} \delta(r_{ij} - r), \tag{1.42}$$

where $r_{ij} \equiv |\mathbf{r}_i - \mathbf{r}_j|$ is the distance between two particles, while $r > 0$. The pair density function is defined as the expectation value of the pair density operator, that is

$$\rho_2^{(N)}(r) \equiv \langle \hat{\rho}_2^{(N)}(r) \rangle. \tag{1.43}$$

The density function at short distances ($r \to 0$) is estimated as follows

$$\begin{aligned}
\rho_2^{(N)}(r) &= \langle \psi | \hat{\rho}_2^{(N)}(r) | \psi \rangle = \int d\mathbf{r}_1 \cdots d\mathbf{r}_N \frac{1}{r^2} \sum_{i<j=1}^{N} \delta(r_{ij} - r) |\psi(\mathbf{r}_1,\ldots,\mathbf{r}_N)|^2 \\
&= \frac{N(N-1)}{2} \int d\mathbf{r}_{12} \frac{1}{r^2} \delta(r_{12} - r) \int \prod_{k=3}^{N} d\mathbf{r}_k d\mathbf{R}_{12} \, |\psi(\mathbf{r}_1,\ldots,\mathbf{r}_N)|^2 \\
&\to \frac{N(N-1)}{2} \int d\mathbf{r}_{12} \frac{1}{r^2} \delta(r_{12} - r) |\phi_2(\mathbf{r}_{12})|^2 \\
&\quad \int \prod_{k=3}^{N} d\mathbf{r}_k d\mathbf{R}_{12} \, |A_2^{(N)}(\mathbf{R}_{12}; \{\mathbf{r}_k\}_{k=3}^{N})|^2 \\
&= C_2^{(N)} \int d\Omega \, |\phi_2(\mathbf{r})|^2.
\end{aligned} \tag{1.44}$$

Defining $\rho_2(r) \equiv \int d\Omega \, |\phi_2(\mathbf{r})|^2$, we obtain the short-distance asymptotics

$$\rho_2^{(N)}(r) \to C_2^{(N)} \rho_2(r), \quad r \to 0. \tag{1.45}$$

We already have our first short-distance asymptotics which, as promised, relates known quantities—the contact and the two-body ground state—to a many-body observable at short distances.

**Expectation value of the interaction.** This is the simplest of relations. The expectation value of any of the two-body interactions, if this has a short range, in the many-body state is simply

$$\langle\psi|V(\mathbf{r}_i - \mathbf{r}_j)|\psi\rangle = \int d\mathbf{r}\rho_2^{(N)}(r)V(r) = C_2^{(N)}\int d\mathbf{r}\rho_2(r)V(r)$$
$$= C_2^{(N)}\langle\phi_2|V|\phi_2\rangle. \tag{1.46}$$

**High-momentum limit of the momentum distribution.** Probing short distances is essentially equivalent to probing high momenta in a quantum many-body system. Therefore, we take a look now at the one-body momentum distribution $\rho(\mathbf{k})$ of the many-boson system. The simplest expression for $\rho(\mathbf{k})$ to work with is given by

$$\rho(\mathbf{k}) = N\int d\mathbf{r}_2 \ldots d\mathbf{r}_N \left|\int d\mathbf{r}_1 e^{-i\mathbf{k}\cdot\mathbf{r}_1}\psi(\mathbf{r}_1,\ldots,\mathbf{r}_N)\right|^2$$
$$\equiv N\int d\mathbf{r}_2 \ldots d\mathbf{r}_N \, |I(\mathbf{k}; \{\mathbf{r}_j\}_{j=2}^N)|^2. \tag{1.47}$$

At high momenta, only the short-range parts, with respect to $\mathbf{r}_1$, contribute significantly to the integral $I$. Hence, as $\mathbf{k} \to \infty$,

$$I(\mathbf{k}; \{\mathbf{r}_j\}_{j=2}^N) \to \sum_{j=2}^N \int d\mathbf{r}_1 e^{-i\mathbf{k}\cdot(\mathbf{r}_1 - \mathbf{r}_j)}e^{-i\mathbf{k}\cdot\mathbf{r}_j}\phi_2(\mathbf{r}_{1j})A_2^{(N)}(\mathbf{R}_{1j}; \{\mathbf{r}_k\}_{k\neq 1,j})$$
$$= \sum_{j=2}^N \int d\mathbf{r}_{1j} e^{-i\mathbf{k}\cdot\mathbf{r}_{1j}}\phi_2(\mathbf{r}_{1j})A_2^{(N)}(\mathbf{R}_{1j}; \{\mathbf{r}_k\}_{k\neq 1,j}) \tag{1.48}$$
$$= \tilde{\phi}_2(\mathbf{k})\sum_{j=2}^N e^{-i\mathbf{k}\cdot\mathbf{r}_j}A_2^{(N)}(\mathbf{R}_{1j}; \{\mathbf{r}_k\}_{k\neq 1,j}),$$

where we have defined the Fourier transform of the two-body wave function

$$\tilde{\phi}_2(\mathbf{k}) = \int d\mathbf{r}\phi_2(\mathbf{r})e^{-i\mathbf{k}\cdot\mathbf{r}}. \tag{1.49}$$

We now need the squared modulus of the asymptotic integral $I$ in equation (1.48), which contains two contributions, $|I|^2 \to |\tilde{\phi}_2(\mathbf{k})|^2(J_1 + J_2)$, with

$$J_1 = \sum_{j=2}^N \left|A_2^{(N)}(\mathbf{R}_{1j}; \{\mathbf{r}_k\}_{k\neq 1,j})\right|^2, \tag{1.50}$$

and

$$J_2 = \sum_{j\neq\ell=2}^N \left[A_2^{(N)}(\mathbf{R}_{1j}; \{\mathbf{r}_k\}_{k\neq 1,j})\right]^* A_2^{(N)}(\mathbf{R}_{1\ell}; \{\mathbf{r}_k\}_{k\neq 1,\ell})e^{i\mathbf{k}\cdot\mathbf{r}_{j\ell}}. \tag{1.51}$$

For large $\mathbf{k}$, $J_2$ oscillates rapidly and the space integrals in equation (1.47) are negligibly small. Therefore, only $J_1$ contributes in this limit to the momentum distribution, and we finally obtain

$$\rho(\mathbf{k}) \to |\tilde{\phi}_2(\mathbf{k})|^2 N(N - 1) \int d\mathbf{r}_2 \ldots d\mathbf{r}_N \left|A_2^{(N)}(\mathbf{R}_{12}; \{\mathbf{r}_k\}_{k=3}^N)\right|^2, \quad \mathbf{k} \to \infty. \tag{1.52}$$

The above integral relates the large-momentum tails of the momentum distribution with the contact as

$$\rho(\mathbf{k}) \rightarrow 2C_2^{(N)} |\tilde{\phi}_2(\mathbf{k})|^2, \quad \mathbf{k} \rightarrow \infty. \tag{1.53}$$

**Static structure factor.** The static structure factor $S(\mathbf{k})$ is another relevant physical quantity that can be experimentally probed. In fact, for $^4$He, neutron scattering experiments, which measure $S(\mathbf{k})$ up to momenta as large as 70 nm$^{-1}$, have been available for over 40 years [3]. These large momenta allow for probing the inner structure of $^4$He droplets and liquids at low temperature, where the typical short-distance length scale is about 0.5 nm, see table 1.1. This fact, together with the availability of very precise model potentials for $^4$He–$^4$He interactions, allow us direct experimental verification of the large-momentum tails of the structure factor. In this case, the analysis is straightforward, and for spherically symmetric interactions, we obtain

$$S(\mathbf{k}) \rightarrow 1 + \frac{8\pi C_2^{(N)}}{Nk} \int_0^\infty dr\, r \sin(kr)\rho_2(r), \tag{1.54}$$

where $\rho_2(r)$ is defined right above equation (1.45).

We analyze now the contact using experimental data. We begin by extracting the contact for $^4$He at a temperature of 1.1 K, using the pair correlation function from reference [3] and the probability density for the helium dimer from reference [1]. The contact can be obtained using equation (1.45), which gives, in this case, $C_2^{(N)}/N \approx 210$ in the thermodynamic limit, in agreement with the diffusion Monte Carlo simulations of reference [4], that reported a value of $C_2^{(N)}/N = 230 \pm 25$. The comparison of equation (1.45) and the pair correlation function is shown in figure 1.2.

Before taking a look at the experimental data for the static structure factor of liquid $^4$He, we can analyze the general form of the asymptotic behaviour using the zero-energy asymptotic wave function. Since $\bar{u}(r) \propto r - (a + a_1 r^{-3} + a_2 r^{-4} + a_3 r^{-5})$, then the integrand in equation (1.54) contains the term

$$r\,|R(r)|^2 \rightarrow \frac{|\bar{u}(r)|^2}{r} \propto \frac{(r-a)^2}{r} + \sum_{n \geqslant 3} \frac{\alpha_n}{r^n}, \tag{1.55}$$

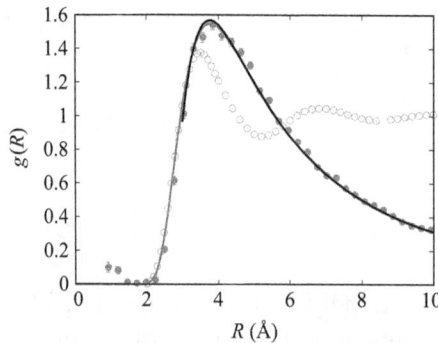

**Figure 1.2.** Pair correlation $g(R)$ as a function of distance. Experimental pair correlation function (open dots) of liquid $^4$He at $T = 1$ K, from reference [3]. Red dots are the normalized two-body pair density function, from reference [1], rescaled by the contact $C^{(N)}/N = 210$. Solid lines are the same but for the zero-energy state with the Aziz potential, both exact (blue) and asymptotic (black).

where the first term on the right-hand-side corresponds with the zero-range limit, and the $\alpha_n$ are coefficients, whose particular forms are not of much relevance. For large $k$, the part containing the zero-range limit gives a contribution to $S(\mathbf{k})$ that is proportional to $k^{-1}$. This, however, cannot be the case, since the wave function at very short distances is essentially zero. The rest of the Fourier transforms must be evaluated with a short-distance cutoff $r_c$, since they are too divergent as $r \to 0$. This is, in any case, natural, since the HE–He interaction contains a short-distance hard core, beyond which the wave functions vanish, around $r_c \approx 2.5$ Å. Therefore, we evaluate the integrals

$$I(n, k) = \frac{1}{k} \int_{r_c}^{\infty} dr\, r \sin(kr) \frac{1}{r^n} = k^{n-3} \int_{kr_c}^{\infty} dz \frac{\sin z}{z^{n-1}}, \quad n \geqslant 3. \tag{1.56}$$

The leading terms are given by

$$I(3, k) = \frac{\cos(kr_c)}{(kr_c)^2} + O((kr_c)^3) \tag{1.57}$$

$$I(4, k) = 3r_c \frac{\sin(kr_c)}{(kr_c)^3} + O((kr_c)^4). \tag{1.58}$$

If we only keep the leading-order terms, we see that the structure factor should show damped oscillations with period $2\pi/r_c$. The 'zero-range' term, if we include the short-distance cutoff, also contributes as $\propto \cos(kr_c)/(kr_c)^2$. Therefore, the structure factor, at large $k$, in this approximation, behaves as

$$S(\mathbf{k}) \to 1 + \frac{8\pi C_2^{(N)}}{N}(a^2 + \alpha_3) \frac{\cos(kr_c)}{(kr_c)^2} + O((kr_c)^{-3}). \tag{1.59}$$

This essentially means that, since in this approximation we do not have information about $r \to 0$, there is a range of momenta, large but not extending to infinity, in which the structure factor behaves as in equation (1.59). Indeed, we see in figure 1.3 that, for $k$ in the range $[2.5, 5]$ Å$^{-1}$, the structure factor is essentially an oscillating function. Beyond that point, we need the actual short-distance two-body wave function.

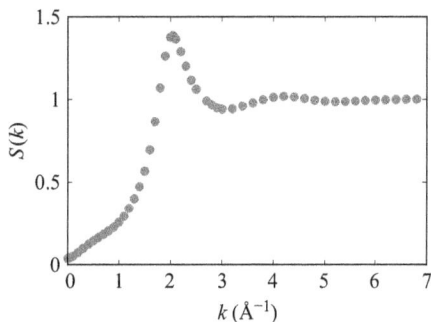

**Figure 1.3.** Static structure factor for liquid helium, with data from the experiments in reference [3].

### 1.3.2 Spin-1/2 fermions

We study now short-range correlations for spin-1/2 fermions. These were the systems originally studied by Tan, resulting in the so-called Tan's relations, in the zero-range limit, valid for ultracold atomic systems [5–7]. We begin, however, with weak universality, and connect with Tan's and subsequent work in the following section.

A many-body eigenstate for spin-1/2 fermions, in the first quantization and in the position representation, has the form $\psi(\mathbf{r}_1\sigma_1, \mathbf{r}_2\sigma_2,...,\mathbf{r}_N\sigma_N)$, where $\sigma_i$ $(i = 1, 2,...,N)$ is the $z$-projection of the spin of particle $i$. As such, we will denote $\sigma_i = 1/2 \equiv \uparrow$, and $\sigma_i = -1/2 \equiv \downarrow$, and call these spin-up and spin-down, respectively. In general, if there are no spin flipping terms in the Hamiltonian, the total spin projection is conserved. When this is the case, $N = N_\uparrow + N_\downarrow$, with $N_\uparrow$ the conserved number of spin-up particles, and equivalently for $N_\downarrow$, the number of spin-down particles. Before proceeding, we shall call the two-body ground state for antiparallel spins in the relative coordinate $\phi_{\uparrow\downarrow}(\mathbf{r})$, which is typically a singlet (unless, for instance, there are external magnetic fields). If the Lagrangian is SU(2)-symmetric, then the two-body ground state in the parallel-spin sector is unique and given by $\phi_{\uparrow\uparrow}(\mathbf{r}) = \phi_{\downarrow\downarrow}(\mathbf{r})$. In the many-body problem, we will choose the first $N_\uparrow$ particle labels to correspond to spin-up fermions, and $i = N_\uparrow + 1,...,N$ to spin-down fermions.

We consider the many-body system to evolve according to the Schrödinger equation with Hamiltonian (1.37), which is independent of spin. When two particles $i$ and $j$ approach each other, the wave function is factorized as

$$\psi(\mathbf{r}_1 \uparrow,...,\mathbf{r}_N \downarrow) \to \phi_{\sigma_i\sigma_j}(\mathbf{r}_{ij})A_{\sigma_i\sigma_j}(\mathbf{R}_{ij}; \{\mathbf{r}_k\}_{k\neq i,j}), \quad \mathbf{r}_{ij} \to 0. \tag{1.60}$$

We define now three different contacts, namely $C_{\uparrow\downarrow}$, $C_{\uparrow\uparrow}$ and $C_{\downarrow\downarrow}$. Note that, in general, if $N_\uparrow \neq N_\downarrow$, it will hold that $C_{\uparrow\uparrow} \neq C_{\downarrow\downarrow}$. The definitions are as follows

$$C_{\uparrow\downarrow} = N_\uparrow N_\downarrow \langle A_{\uparrow\downarrow}|A_{\uparrow\downarrow}\rangle, \tag{1.61}$$

$$C_{\sigma\sigma} = \frac{N_\sigma(N_\sigma - 1)}{2}\langle A_{\sigma\sigma}|A_{\sigma\sigma}\rangle, \quad \sigma = \uparrow, \downarrow, \tag{1.62}$$

where the norms of the residual wave functions $A_{\sigma\sigma'}$ are defined as in equation (1.40). We proceed now to study short-range correlations in the different spin channels.

**Pair density function.** The pair density operator, equation (1.42), can be split into three terms, namely $\hat{\rho}_2^{(N)}(r) \equiv \hat{\rho}_{\uparrow\downarrow}(r) + \hat{\rho}_{\uparrow\uparrow}(r) + \hat{\rho}_{\downarrow\downarrow}(r)$. Explicitly, these are defined as

$$\hat{\rho}_{\uparrow\downarrow}(r) = \frac{1}{r^2}\sum_{i=1}^{N_\uparrow} \sum_{j=N_\uparrow+1}^{N} \delta(r_{ij} - r), \tag{1.63}$$

$$\hat{\rho}_{\uparrow\uparrow}(r) = \frac{1}{r^2}\sum_{i<j=1}^{N_\uparrow} \delta(r_{ij} - r), \tag{1.64}$$

$$\hat{\rho}_{\downarrow\downarrow}(r) = \frac{1}{r^2} \sum_{i<j=1}^{N_\uparrow+1} \delta(r_{ij} - r) \tag{1.65}$$

Following the analysis done in the previous section, equation (1.44), we arrive at similar expressions for the short-range limit of the two-body density functions $\rho_2^{(N)}(r) = \langle \hat{\rho}_2^{(N)}(r) \rangle = \rho_{\uparrow\downarrow}(r) + \rho_{\uparrow\uparrow}(r) + \rho_{\downarrow\downarrow}(r)$, as

$$\rho_2^{(N)}(r) \rightarrow C_{\uparrow\downarrow}\rho_2^{\uparrow\downarrow}(r) + C_{\uparrow\uparrow}\rho_2^{\uparrow\uparrow}(r) + C_{\downarrow\downarrow}\rho_2^{\downarrow\downarrow}(r), \quad r \rightarrow 0. \tag{1.66}$$

where the two-body densities for two particles are defined analogously to the bosonic case, as

$$\rho_2^{\sigma\sigma'}(r) \equiv \int d\Omega \, |\phi_{\sigma\sigma'}(\mathbf{r})|^2. \tag{1.67}$$

**Expectation value of the interaction.** For interactions with SU(2) symmetry, we immediately obtain

$$\frac{1}{\binom{N}{2}} \sum_{i<j=1}^{N} \langle V(\mathbf{r}_i - \mathbf{r}_j) \rangle = \frac{2N_\uparrow N_\downarrow}{N(N-1)} C_{\uparrow\downarrow}\langle \phi_{\uparrow\downarrow}|V|\phi_{\uparrow\downarrow}\rangle + \frac{N_\uparrow(N_\uparrow - 1)}{N(N-1)} C_{\uparrow\uparrow}\langle \phi_{\uparrow\uparrow}|V|\phi_{\uparrow\uparrow}\rangle \tag{1.68}$$

$$+ \frac{N_\downarrow(N_\downarrow - 1)}{N(N-1)} C_{\downarrow\downarrow}\langle \phi_{\downarrow\downarrow}|V|\phi_{\downarrow\downarrow}\rangle.$$

For a spin-balanced system ($N_\uparrow = N_\downarrow = N/2$), and in the thermodynamic limit ($N \rightarrow \infty$), we have that $C_{\uparrow\uparrow} = C_{\downarrow\downarrow}$ and the result simplifies to

$$\frac{1}{\binom{N}{2}} \sum_{i<j=1}^{N} \langle V(\mathbf{r}_i - \mathbf{r}_j) \rangle = \frac{1}{2}[C_{\uparrow\downarrow}\langle \phi_{\uparrow\downarrow}|V|\phi_{\uparrow\downarrow}\rangle + C_{\uparrow\uparrow}\langle \phi_{\uparrow\uparrow}|V|\phi_{\uparrow\uparrow}\rangle], \quad N_\uparrow = N_\downarrow \rightarrow \infty. \tag{1.69}$$

**High momentum limit of the momentum distribution.** For the large-$k$ limit of the momentum distribution, we proceed once more in analogy with the bosonic case of the previous section. We separate the momentum distribution into the two spin components. Following our convention, namely that for labels $i = 1,...,N_\uparrow$ the particles have spin-up, and the rest have spin-down, we define

$$\rho_\uparrow(\mathbf{k}) = N_\uparrow \int d\mathbf{r}_2 \dots d\mathbf{r}_N \left| \int d\mathbf{r}_1 e^{-i\mathbf{k}\cdot\mathbf{r}_1}\psi(\mathbf{r}_1,...,\mathbf{r}_N) \right|^2$$

$$\equiv N_\uparrow \int d\mathbf{r}_2 \dots d\mathbf{r}_N \, |I_\uparrow(\mathbf{k}; \{\mathbf{r}_j\}_{j=2}^N)|^2, \tag{1.70}$$

$$\rho_\downarrow(\mathbf{k}) = N_\downarrow \int d\mathbf{r}_1 \cdots d\mathbf{r}_{N-1} \left| \int d\mathbf{r}_N e^{-i\mathbf{k}\cdot\mathbf{r}_N}\psi(\mathbf{r}_1,...,\mathbf{r}_N) \right|^2$$

$$\equiv N_\downarrow \int d\mathbf{r}_1 \cdots d\mathbf{r}_{N-1} \, |I_\downarrow(\mathbf{k}; \{\mathbf{r}_j\}_{j=1}^{N-1})|^2, \tag{1.71}$$

The development of the integrals $I_\uparrow$ and $I_\downarrow$ parallels the bosonic case. The final result reads

$$\rho_\uparrow(\mathbf{k}) \to C_{\uparrow\downarrow} \, |\tilde{\phi}_{\uparrow\downarrow}(\mathbf{k})|^2 + 2C_{\uparrow\uparrow} \, |\tilde{\phi}_{\uparrow\uparrow}(\mathbf{k})|^2, \quad \mathbf{k} \to \infty, \tag{1.72}$$

and analogously, replacing $\uparrow$ with $\downarrow$ and viceversa, for $\rho_\downarrow(\mathbf{k})$.

**Static structure factor.** Once more, the derivation of the large-$k$ asymptotics of the static structure factor $S(\mathbf{k})$ parallels the bosonic case. It can be written as $S(\mathbf{k}) = 1 + \Sigma_{\uparrow\downarrow}(\mathbf{k}) + \Sigma_{\uparrow\uparrow}(\mathbf{k}) + \Sigma_{\downarrow\downarrow}(\mathbf{k})$, with

$$\Sigma_{\sigma\sigma'}(\mathbf{k}) \to \frac{8\pi C_{\sigma\sigma'}}{Nk} \int_0^\infty dr\, r \sin(kr) \rho_2^{\sigma\sigma'}(r), \quad \mathbf{k} \to \infty. \tag{1.73}$$

This analysis finishes the weak universal regime for two of the most interesting, and typical systems in low-temperature physics. Next, we study the strong universal regime, and derive a number of non-perturbative results, known as Tan's relations, in the zero-range limit.

## 1.4 Strong short-range universality: Tan's relations

In this section, we explore strong universality at short distances. To see what we mean by this, we need to refer to section 1.2, where we discussed the relevant length scales for a few examples of experimental relevance. The case of liquid $^4$He, which we studied as the main example for weak universality, was quite clear: the full two-body ground state is necessary to study short-range asymptotics in a number of quantities, due to the comparable interaction range and interparticle distance (see table 1.1). If, instead, we look up the fermionic $^{40}$K in table 1.1, we see that while the interaction range is of the order of a few nanometers, the interparticle distance is of the order of $10^5$ nm. The separation of these two length scales is so large that the non-universal, interaction-dependent part of the two-body wave function will only be of relevance in the distance range $r \leqslant r_c \approx 0.5$ Å. Oscillations, which naturally appear in the structure factor, will not be appreciable unless momenta $k \gtrsim r_c^{-1}$ can be probed. In the case of liquid $^4$He, this meant $k \gtrsim 0.3$ Å$^{-1}$, that is, damped oscillations should be (and are!) visible and important at all momentum scales, see figure 1.3. In the case of $^{40}$K, oscillations will play a role for $k \gtrsim 2$ Å$^{-1}$. State-of-the-art experiments (see reference [8]), probe the structure factor at momenta $k$ satisfying $k_F/k > 0.1$. From table 1.1, $k_F \approx 2 \times 10^{-7}$ Å$^{-1}$, and therefore the probed momenta are bounded from above as $k < 2 \times 10^{-6}$ Å$^{-1}$. While this is a large momentum for the system at hand, it is extremely small when compared with the non-universal regime appearing at $k > 2$ Å$^{-1}$. The situation in experiments using $^6$Li, is very similar, with $k_F \sim 10^{-4}$ Å$^{-1}$. Hence, strong universality, where the details of the short-range part of the interaction are meaningless, will emerge.

Luckily, strong universality does not require much more technology than what was done for its weak version. More importantly, in its strong form, short-range universality allows for a number of universal, non-perturbative relations that are unavailable for weakly-universal systems. We consider a system of non-relativistic spin-1/2 fermions interacting via the $s$-wave leading-order interaction in effective field theory (low-energy (LO) interaction in EFT, see volume 1) in three spatial dimensions. In Hamiltonian form, and in the position representation, we choose the Huang–Yang pseudopotential for the interaction, that is

$$H = -\frac{\hbar^2}{2m}\sum_{i=1}^{N} \nabla_i^2 + g_R \sum_{i<j=1}^{N} \delta(\mathbf{r}_{ij})\frac{\partial}{\partial r_{ij}}(r_{ij}\cdot), \tag{1.74}$$

where $g_R = 4\pi\hbar^2 a/m$ is the renormalized LO EFT coupling constant, and $a$ is the $s$-wave scattering length. As a simple warning, the argument of the Dirac delta function in the Huang–Yang pseudopotential, equation (1.74), must be a vector, not a scalar distance. Since the interaction has zero range, and is $s$-wave, only fermions with opposite spin interact. The two-particle lowest-energy state is a singlet, with spatial wave function

$$\phi_{\uparrow\downarrow}(\mathbf{r}) = R(r)Y_{00}(\Omega), \tag{1.75}$$

and

$$R(r) = \sqrt{\frac{2}{a}}\frac{e^{-r/a}}{r}, \quad a > 0, \tag{1.76}$$

$$R(r) = \frac{1}{r} - \frac{1}{a}, \quad a < 0. \tag{1.77}$$

For $a > 0$, the zero-energy state is also given by equation (1.77). Atomic gases, when cooled below the degeneracy temperature, and probed during time scales much shorter than molecular formation, remain atomic. That is, regardless of whether $a > 0$ (effectively *repulsive*) or $a < 0$ (effectively weakly attractive), the many-body system remains a gas, and does not feature deeply bound molecules, which would correspond to the ground state. These, in any case, are lost from the system when formed since the external traps do not interfere meaningfully with the molecular level structure. Therefore, for interparticle separations $R_{\text{vdW}} \ll r \ll R_*$, we may use $R(r)$ in equation (1.77) as the short-range limit of the two-body wave function. We will also explore corrections due to the tails, and see whether these are relevant or not for the problem at hand.

Without further ado, we delve into the derivation of the strong short-range universal relations for a spin-1/2 Fermi gas. We offer a straightforward adaptation of the method used for the weak universal systems.

### 1.4.1 Short-range asymptotics and Tan's contact

Here, we do not have much to do, except for deriving the universal relations themselves, and linking the definition of the contact for weakly-universal systems to the definition for strongly universal systems.

We begin with equation (1.60) and, instead of the normalized two-body ground state, we use the (infinite size) zero-energy singlet state, with the convention

$$\phi_{\uparrow\downarrow}(\mathbf{r}) = \left( \frac{1}{r} - \frac{1}{a} \right) Y_{00}(\Omega). \tag{1.78}$$

This convention, in fact, corresponds to Dirac delta normalization of scattering states (see volume 1), $\langle \mathbf{k}'|\mathbf{k} \rangle = (2\pi)^3 \delta(\mathbf{k} - \mathbf{k}')$, but in the zero-energy limit, and in the $s$-wave channel. Since the scattering operator is unitary, normalization does not change upon collisions. The $s$-wave contribution to a three-dimensional plane wave $\exp(i\mathbf{k} \cdot \mathbf{r})$ is simply $j_0(kr) = \sin(kr)/kr$ which, in the zero-energy limit, is simply 1. Using the orthogonality relation for spherical Bessel functions,

$$\int_0^\infty dr r^2 j_0(k'r) j_0(kr) = \frac{\pi}{2k^2}\delta(k - k') = 2\pi^2\delta(\mathbf{k} - \mathbf{k}'), \tag{1.79}$$

the two-body zero-energy state in equation (1.78) is normalized to $2\pi^2\delta(\mathbf{k} - \mathbf{k}')$ as $\mathbf{k} \to 0$. To have a properly normalized state, with normalization $\langle \phi_{\uparrow\downarrow}^{k'}|\phi_{\uparrow\downarrow}^{k} \rangle = (2\pi)^3\delta(\mathbf{k} - \mathbf{k}')$, we must multiply equation (1.78) by $\sqrt{4\pi} = 1/Y_{00}(\Omega)$. For parallel spins, we use their $p$-wave zero-energy state, which is non-interacting, and given by

$$\phi_{\sigma\sigma}(r) \propto \frac{j_1(kr)}{k} \propto r, \quad k \to 0. \tag{1.80}$$

In this case, we shall not care about normalization since the interaction is purely $s$-wave, and there will not be a contact $C_{\sigma\sigma}$.

We write the many-fermion wave function in the second quantization using creation ($\hat{\Psi}_\sigma^\dagger(\mathbf{r})$) and annihilation ($\hat{\Psi}_\sigma(\mathbf{r})$) operators, satisfying canonical anticommutation relations,

$$\{\hat{\Psi}_\sigma(\mathbf{r}), \hat{\Psi}_{\sigma'}^\dagger(\mathbf{r}')\} = \delta_{\sigma\sigma'}\delta(\mathbf{r} - \mathbf{r}'). \tag{1.81}$$

The wave function is then given by

$$|\psi\rangle = \frac{1}{N_\uparrow! N_\downarrow!} \int d\mathbf{r}_1 \cdots d\mathbf{r}_N \psi(\mathbf{r}_1,...,\mathbf{r}_N) \prod_{j=1}^{N_\uparrow} \hat{\Psi}_\uparrow^\dagger(\mathbf{r}_j) \prod_{l=N_\uparrow+1}^{N} \hat{\Psi}_\downarrow^\dagger(\mathbf{r}_l)|0\rangle. \tag{1.82}$$

The normalization of the full many-fermion wave function $\langle \psi|\psi \rangle = 1$ is guaranteed with the condition

$$\frac{1}{N_\uparrow! N_\downarrow!} \int d\mathbf{r}_1,...\mathbf{r}_N \, |\psi(\mathbf{r}_1,...,\mathbf{r}_N)|^2 = 1. \tag{1.83}$$

We now consider two particles ($i \leqslant N_\uparrow$ and $j > N_\uparrow$) with opposite spins in close proximity in comparison with all other interparticle distances. The wave function is simply

$$\psi(\mathbf{r}_1,...,\mathbf{r}_N) \to \left( \frac{1}{r_{ij}} - \frac{1}{a} \right) A_{\uparrow\downarrow}(\mathbf{R}_{ij}; \{\mathbf{r}_k\}_{k\neq i,j}), \quad \mathbf{r}_{ij} \to 0. \tag{1.84}$$

We will define the contact, and call it Tan's contact, as

$$C_{\uparrow\downarrow} \equiv (4\pi)^2 N_\uparrow N_\downarrow \langle A_{\uparrow\downarrow} | A_{\uparrow\downarrow} \rangle. \tag{1.85}$$

Note that the only difference in convention here with respect to our analysis of the weak universal regime is a $(4\pi)^2$ factor that will prove convenient later on. With these definitions, almost everything proceeds as in the weakly-universal case.

**Pair density function.** The pair density function at short distances behaves as

$$\rho_2^{(N)}(r) \rightarrow \frac{C_{\uparrow\downarrow}}{4\pi} \left( \frac{1}{r} - \frac{1}{a} \right)^2, \quad r \rightarrow 0. \tag{1.86}$$

**High momentum limit of the momentum distribution.** The momentum distribution for spin-$\sigma$ fermions is given, at large $k$, by

$$\rho_\sigma(\mathbf{k}) \rightarrow \frac{C_{\uparrow\downarrow}}{k^4}, \quad \mathbf{k} \rightarrow \infty, \tag{1.87}$$

where we have used the Fourier transform

$$\int d\mathbf{r} e^{i\mathbf{k}\cdot\mathbf{r}} \frac{1}{r} = \frac{4\pi}{k^2}. \tag{1.88}$$

**Static structure factor.** The large momentum behaviour of $\Sigma_{\uparrow\downarrow}(\mathbf{k})$ is in this case

$$\Sigma_{\uparrow\downarrow}(\mathbf{k}) \rightarrow \frac{C_{\uparrow\downarrow}}{N} \left[ \frac{1}{8k} - \frac{1}{2\pi a k^2} \right]. \tag{1.89}$$

which includes only the divergent part (as $r \rightarrow 0$) of the two-body density.

### 1.4.2 Tan's universal relations.

Recall that, in the previous subsection, we omitted, as a relation, the expectation value of the interaction in the zero-range approximation. This is because it features an ultraviolet (UV) divergence which, of course, cancels out with another UV divergence that occurs in the expectation value of the kinetic energy. This is simple to see, since at large momenta $\rho_\sigma(\mathbf{k}) \propto k^{-4}$, and therefore, placing a UV cutoff $\Lambda$,

$$\left\langle \frac{\hbar^2 k^2}{2m} \right\rangle = \frac{2\pi\hbar^2}{m} \int_0^\Lambda dk k^4 [\rho_\uparrow(k) + \rho_\downarrow(k)] = \frac{4\pi\hbar^2 C_{\uparrow\downarrow}}{m} \Lambda + O(N). \tag{1.90}$$

As expected, the kinetic energy is UV-divergent. The total energy of the system is, of course, well-defined, because the LO EFT interaction renormalizes the problem, yielding a finite energy. But the interaction energy must be clearly divergent with a leading-order term of the form $-4\pi\hbar^2 C_{\uparrow\downarrow}\Lambda/m$. Therefore, our first goal is to derive the first of Tan's relations, namely the energy relation, before proceeding with the rest.

**Energy relation.** The expectation value of the total energy of the system $\langle H \rangle = E$ is given by

$$E = \frac{2\pi\hbar^2}{m} \int_0^\infty dk k^4 \left[ \rho_\uparrow(k) + \rho_\downarrow(k) - \frac{2C_{\uparrow\downarrow}}{k^4} \right] + \frac{\hbar^2}{4\pi ma} C_{\uparrow\downarrow}. \tag{1.91}$$

Note that the contact term in the integrand above, scaling as $1/k^4$, cancels out the UV divergence in the kinetic energy and, therefore, the total energy is finite (of $O(N)$).

**Density–density correlation.** The two-body distribution function for spin-antiparallel fermions, is given by equation (A.5) in appendix A so, at short distances, $g_{\uparrow\downarrow}(\mathbf{r})$ takes the form

$$g_{\uparrow\downarrow}(\mathbf{r}) \to \frac{C_{\uparrow\downarrow}}{16\pi^2} \left( \frac{1}{r^2} - \frac{2}{ar} \right). \tag{1.92}$$

**Adiabatic relation.** Changes in the scattering length imply, of course, changes in the energy. These are given by

$$\frac{dE}{d(1/a)} = -\frac{\hbar^2}{4\pi m} C_{\uparrow\downarrow}. \tag{1.93}$$

This is known as Tan's adiabatic relation, or adiabatic theorem.

**Pressure relation.** For a homogeneous system, the pressure $P$ and the energy density $\mathcal{E} = E/V$ are related via

$$P = \frac{2}{3}\mathcal{E} + \frac{\hbar^2}{12\pi ma V} C_{\uparrow\downarrow}. \tag{1.94}$$

Above, $V$ is the volume of the system.

**Generalized virial theorem.** If the Fermi gas is trapped by an external harmonic potential $W \propto \mathbf{r}^2$, then the energy satisfies the following generalization of the virial theorem

$$E = 2\langle W \rangle - \frac{\hbar^2}{8\pi ma} C_{\uparrow\downarrow}. \tag{1.95}$$

**Proofs of Tan's relations.** We now prove all the universal relations we have just stated. We try to use a variety of methods of proof, but note that these are provable in several other ways (see, e.g. reference [9]).

We begin with the energy theorem. To show it, we consider the expectation value of the kinetic energy $\langle T \rangle$, using the position representation, which is reduced to

$$\langle T \rangle = -\frac{\hbar^2}{2m} \left[ N_\uparrow \langle \nabla^2_{\mathbf{r}_1} \rangle + N_\downarrow \langle \nabla^2_{\mathbf{r}_N} \rangle \right]. \tag{1.96}$$

We are interested in the minimal divergent part of the kinetic energy only. We set a short-distance cutoff $r_0$ in the integrals involved, and we obtain

$$\langle \nabla^2_{\mathbf{r}_1} \rangle = \sum_{j=N_\uparrow+1}^{N} \int d\mathbf{r}_1 \cdots d\mathbf{r}_N |A_{\uparrow\downarrow}(\mathbf{R}_{1j}; \{\mathbf{r}_l\}_{l\neq 1,j})|^2 \frac{1}{r_{1j}^4} + \text{Regular terms.} \tag{1.97}$$

We get a completely analogous expression for $\langle \nabla^2_{\mathbf{r}_N} \rangle$. The short-range part of the kinetic energy $\langle T \rangle_{SR}$ therefore reads

$$\langle T \rangle_{SR} = \frac{\hbar^2 C_{\uparrow\downarrow}}{4\pi m r_0}, \quad r_0 \to 0. \tag{1.98}$$

We perform the same analysis for the interaction energy, which can be obtained in closed form, and add the short-distance cutoff in the delta interaction as $\delta(r_{ij} - r_0)$. We obtain the full interaction energy as

$$\langle V \rangle = -\frac{\hbar^2 C_{\uparrow\downarrow}}{4\pi m r_0} + \frac{\hbar^2 C_{\uparrow\downarrow}}{4\pi m a}. \tag{1.99}$$

Therefore, the energy has the form

$$E = \int d\mathbf{k} \frac{\hbar^2 k^2}{2m} \left[ \rho_\uparrow(\mathbf{k}) + \rho_\downarrow(\mathbf{k}) - \frac{2C_{\uparrow\downarrow}}{k^4} \right] + \frac{\hbar^2 C_{\uparrow\downarrow}}{4\pi m a}, \tag{1.100}$$

which proves the result.

To prove the adiabatic theorem, we will use the Hellmann–Feynman theorem, which states, for a regular Hamiltonian that does not contain distributional interactions,

$$\frac{dE}{da} \overset{?}{=} \left\langle \frac{\partial V}{\partial a} \right\rangle. \tag{1.101}$$

For the LO EFT interaction, it is easy to calculate, with a short-distance cutoff $r_0$, the expression on the right-hand-side of equation (1.101), and we obtain

$$\left\langle \frac{\partial V}{\partial a} \right\rangle = -\frac{\hbar^2 C_{\uparrow\downarrow}}{4\pi m a r_0} + \frac{\hbar^2 C_{\uparrow\downarrow}}{4\pi m a^2}, \tag{1.102}$$

which is clearly divergent and, therefore, not correct. It appears obvious that the divergent term should disappear, as it should, since it is divergent. But let us do this rigorously. Firstly, we should wonder why the Hellmann–Feynmann theorem does not work. The reason is very simple: the Hamiltonian is *not* self-adjoint! It is not true that if $H(a)|\psi_a\rangle = E(a)|\psi_a\rangle$ then $\langle \psi_a|H(a) = \langle \psi_a|E(a)$, where we have explicitly included the dependence on the scattering length $a$. To make the divergent term disappear, we should verify that, instead

$$\langle \psi_a|H(a)\partial_a\psi_a\rangle = E(a)\langle \psi_a|\partial_a\psi_a\rangle + \frac{\hbar^2 C_{\uparrow\downarrow}}{4\pi m a r_0}. \tag{1.103}$$

It is a lengthy but straightforward calculation to show that

$$\langle \psi_a|V\partial_a\psi_a\rangle = -\left\langle \frac{\partial V}{\partial a} \right\rangle + \frac{4\pi\hbar^2}{m}\left(\frac{1}{r_0} - \frac{1}{a}\right)N_\uparrow N_\downarrow \langle A_{\uparrow\downarrow}|\partial_a A_{\uparrow\downarrow}\rangle. \tag{1.104}$$

Then, using that $\partial_a \langle \psi_a | \psi_a \rangle = 0$, together with $\langle \psi_a | H(a)^\dagger = E(a) \langle \psi_a |$, we find

$$\frac{\partial E}{\partial a} = -\langle \psi_a | V^\dagger \partial_a \psi_a \rangle - \frac{4\pi\hbar^2}{m} \left( \frac{1}{r_0} - \frac{1}{a} \right) N_\uparrow N_\downarrow \langle A_{\uparrow\downarrow} | \partial_a A_{\uparrow\downarrow} \rangle. \tag{1.105}$$

All that is left is the calculation of $\langle \psi_a | V^\dagger \partial_a \psi_a \rangle = \langle V\psi_a | \partial_a \psi_a \rangle$. Another lengthy calculation gives

$$\langle \psi_a | V^\dagger \partial_a \psi_a \rangle = -\frac{\hbar^2 C_{\uparrow\downarrow}}{4\pi m a^2} - \frac{4\pi\hbar^2}{m} \left( \frac{1}{r_0} - \frac{1}{a} \right) N_\uparrow N_\downarrow \langle A_{\uparrow\downarrow} | \partial_a A_{\uparrow\downarrow} \rangle, \tag{1.106}$$

and, therefore

$$\frac{\partial E}{\partial a} = \frac{\hbar^2 C_{\uparrow\downarrow}}{4\pi m a^2}, \tag{1.107}$$

which finishes the proof of the adiabatic theorem.

As a side note, we can state a variation of the Hellmann–Feynman theorem for EFT, or singular, interactions, which feature non-self-adjointness in intermediate steps. We obtain, in this case, if $H = H(\lambda) = T + V(\lambda)$,

$$\frac{\partial E}{\partial \lambda} = \left\langle \frac{\partial V}{\partial \lambda} \right\rangle + \langle \psi_\lambda | V \partial_\lambda \psi_\lambda \rangle - \langle V\psi_\lambda | \partial_\lambda \psi_\lambda \rangle. \tag{1.108}$$

Obviously, the above generalization to EFT interactions coincides with the standard Hellmann–Feynman theorem for regular interactions. We leave the proof of the above to the interested reader (see problem 3).

Let us prove now the pressure relation. To do that, we fix $N_\uparrow$ and $N_\downarrow$, and calculate $dE/dV$. We prove this relation for a balanced Fermi gas, with $N_\downarrow = N_\uparrow$, and leave the general proof to the problem section. From dimensional analysis, the energy is given by

$$E = \frac{\hbar^2}{ma^2} NF(k_F a), \tag{1.109}$$

where $F$ is a generic continuous, differentiable function (except, perhaps, at $a = 0$), and $k_F$ is the Fermi momentum. The volume $V$ and $k_F$ are related via $V = 3Nk_F^{-3}/4\pi$, and we therefore obtain

$$\frac{dE}{dV} = -\frac{4\pi}{9} \frac{\hbar^2 k_F^4}{ma} F'(k_F a). \tag{1.110}$$

Differentiating the energy with respect to the scattering length, we readily obtain

$$\frac{dE}{da} = -\frac{2}{a} E - \frac{9N}{4\pi k_F^3 a} \frac{dE}{dV}. \tag{1.111}$$

From the above relation, after defining the energy density $\mathcal{E} = E/V$, we obtain

$$\frac{dE}{dV} = -\frac{2}{3}\mathcal{E} - \frac{1}{3V}a\frac{dE}{da}. \tag{1.112}$$

Using the adiabatic theorem, the pressure relation follows.

We add now an external harmonic potential $W \propto \mathbf{r}^2$ to the Hamiltonian, and show the generalized virial theorem. The two length scales are now the scattering length and the oscillator length $\ell = (\hbar/m\omega)^{1/2}$, where $\omega$ is the oscillator's frequency. The energy of the system must depend on these as

$$E = \frac{\hbar^2}{m\ell a}F(\ell/a) = \frac{\hbar^2}{ma^2}G(\ell/a) = \frac{\hbar^2}{m\ell^2}H(\ell/a), \tag{1.113}$$

where $F$, $G$ and $H$ are generic, differentiable functions. It is trivial to verify that all three forms in equation (1.113) are equivalent to one another. We choose the second expression for convenience, and use the Hellmann–Feynman theorem to observe that

$$\frac{\partial E}{\partial \ell} = -\frac{4}{\ell}\langle W \rangle. \tag{1.114}$$

Identifying

$$\frac{dE}{d(1/a)} = 2aE + \frac{\hbar^2\ell}{ma^2}G'(\ell/a), \tag{1.115}$$

we find that

$$\langle W \rangle = \frac{E}{2} - \frac{1}{4a}\frac{dE}{d(1/a)}. \tag{1.116}$$

Using now the adiabatic theorem, we immediately arrive at the generalized virial theorem.

## 1.5 Tan's relations in one spatial dimension

All the universal relations can be adapted to arbitrary dimensions. Since we will devote the entire next chapter to one-dimensional systems with zero-range (to LO or beyond LO EFT) interactions, it is important to derive these relations—although more briefly—in one dimension. For LO EFT interactions, the many-boson problem is called Lieb–Liniger model, the spinless fermionic equivalent (with odd-wave LO EFT interactions) is called Cheon–Shigehara model, while the spin-1/2 fermionic version is called Yang–Yang model. These are all exactly solvable and integrable, which make them perfect playgrounds to test Tan's relations in a controlled way.

### 1.5.1 Lieb–Liniger model

The Lieb–Liniger model consists of $N$ non-relativistic bosons in one dimension interacting via a Dirac delta pairwise potential. That is, their Hamiltonian is given by

$$H = \sum_{i=1}^{N}\frac{p_i^2}{2m} + g\sum_{i<j=1}^{N}\delta(x_i - x_j). \tag{1.117}$$

As we saw in volume 1, the one-dimensional delta potential is regular, and needs no regularization. Given the coupling constant $g$, the one-dimensional scattering length $a_1$ is given by

$$g = -\frac{2\hbar^2}{ma_1}.$$ (1.118)

The short-distance, zero-energy, relative wave function for two particles is given by

$$\phi_2(x_{ij}) = 1 - |x_{ij}|/a_1.$$ (1.119)

At short distances, the many-body wave function $\psi$ separates, as usual, as

$$\psi(x_1,\ldots,x_N) \to (1 - |x_{ij}|/a_1)A_2^{(N)}(X_{ij}; \{x_l\}_{l\neq i,j}), \quad x_{ij} \to 0.$$ (1.120)

where $X_{ij} = (x_i + x_j)/2$ is the centre of mass coordinate for the pair $i$, $j$. We will define the contact as

$$C = N(N - 1)\langle A_2^{(N)}|A_2^{(N)}\rangle,$$ (1.121)

so that, in this case, it is easy to see that it coincides with the integral over the local two-body distribution function,

$$C = \int dx g_2(x, x).$$ (1.122)

Since the Dirac delta interaction in one dimension is regular, the usual Hellmann–Feynman theorem works, and we have

$$\frac{dE}{d(-1/a_1)} = \frac{2\hbar^2}{m}\sum_{i<j}\langle\delta(x_i - x_j)\rangle = \frac{\hbar^2}{m}\int dx g_2(x, x) = \frac{\hbar^2 C}{m}.$$ (1.123)

The energy relation in this case is quite trivial, since $\langle V\rangle = 0$ for $a_1 \to \infty$ (the system becomes non-interacting), and therefore, for finite scattering length, we get, from the adiabatic theorem, $\langle V\rangle = -\hbar^2 C/ma$, and the energy is given by

$$E = \int\frac{dk}{2\pi}\frac{\hbar^2 k^2}{2m}\rho(k) - \frac{\hbar^2 C}{ma}.$$ (1.124)

The tail of the momentum distribution is obtained as usual. All that we need is the distributional Fourier transform of $|x_{ij}|$ which, at large $k$, behaves as $-2/k^2$. Therefore, the large momentum asymptotics of the momentum distribution is

$$\rho(k) \to \frac{4C}{a^2 k^4}, \quad k \to \infty.$$ (1.125)

The structure factor is evaluated, for large momenta, just as in the three-dimensional case, and we have

$$S(k) \to 1 + \frac{4C}{Nak^2}, \quad k \to \infty.$$ (1.126)

For the pressure in a homogeneous system, we recognize the two length scales of the problem as $1/\rho$, with $\rho$ the one-dimensional density, and $a_1$. The energy can be written, from dimensional analysis, as

$$E = \frac{\hbar^2}{ma_1^2} F(\rho a_1).$$

(1.127)

We evaluate $dE/dL$, where $L$ is the length of the system, and use the adiabatic theorem to obtain

$$\frac{\hbar^2 C}{m} = -2a_1 E - a_1 L \frac{dE}{dL},$$

(1.128)

and we obtain, for the pressure $P$,

$$P = 2\mathcal{E} + \frac{\hbar^2 C}{mLa_1},$$

(1.129)

where $\mathcal{E} = E/L$ is the one-dimensional energy density. We leave the one-dimensional bosonic virial theorem to the problem section.

### 1.5.2 Cheon–Shigehara model

The spinless, or spin-polarized fermionic many-body problem with LO odd-wave EFT interactions is more subtle than the Lieb–Liniger model. To begin with, the two-body interaction requires regularization and renormalization (see volume 1). Moreover, for $N > 3$ fermions, the two-body interaction yields further divergences and, therefore, the problem must be once more renormalized. One may resolve this issue in a number of ways: (i) The discrete, nearest-neighbour-coupled version of the model does not require three-body renormalization, has the correct continuum limit, is dual to the Lieb–Liniger model in the continuum, and is exactly solvable via the Bethe ansatz (it is the fermionic version of the Heisenberg model); (ii) we can add a particular three-body interaction or a counterterm in the continuum model and leave the problem exactly solvable; (iii) we can use an associative algebra of generalized functions in order to define products of distributions and in this way define the Cheon–Shigehara model directly as the fermionic dual to the Lieb–Liniger model. This last option will be thoroughly explored in the following section, so we postpone its discussion.

The odd-wave LO interaction can be represented as a boundary condition whenever two fermions approach each other. Let us do this first in the relative coordinate. The wave function must satisfy

$$\psi(0^+) - \psi(0^-) = -2a_1 \psi'(0).$$

(1.130)

The relation above is subtle. The derivative, $\psi'(x)$, which should appear smooth, is in fact a distribution. Therefore, what is meant rigorously by $\psi'(0)$ is

$$\psi'(0) \equiv \lim_{x \to 0^+} \psi'(x) = \lim_{x \to 0^-} \psi'(x),$$

(1.131)

where the derivative must be taken strictly out of the zero limit. We can save ourselves some of this trouble by accepting the distributional nature of the derivative and replacing condition (1.130) with the following integral condition

$$\int_{-\varepsilon}^{\varepsilon} dx \psi'(x) = [\psi(0^+) - \psi(0^-)]\left(1 - \frac{\varepsilon}{a_1}\right), \tag{1.132}$$

where $\varepsilon \to 0$. As such, it is still not so rigorous, so we fix it by differentiating and taking the limit, as

$$\lim_{\varepsilon \to 0^+} \frac{d}{d\varepsilon} \int_{-\varepsilon}^{\varepsilon} dx \psi'(x) = -\frac{1}{a_1}[\psi(0^+) - \psi(0^-)]. \tag{1.133}$$

Equation (1.133) above is the correct zero-range boundary condition, which implies that the zero-energy wave function is given by

$$\psi(x) \propto \mathrm{sgn}(x) - \frac{x}{a_1}. \tag{1.134}$$

Let us calculate the finite-energy scattering states, and the bound state (for $a_1 > 0$), using the boundary condition (1.133). Since we are dealing with fermions with a zero-range interaction, the scattering states have the form (see volume 1)

$$\psi_k(x) = \mathrm{sgn}(x)\sin(k|x| + \theta_k). \tag{1.135}$$

We obtain

$$\lim_{\varepsilon \to 0^+} \frac{d}{d\varepsilon} \int_{-\varepsilon}^{\varepsilon} dx \psi'(x) = 2k \cos \theta_k, \tag{1.136}$$

and applying the boundary condition (1.133), we finally get

$$-k \cot \theta_k = \frac{1}{a_1}. \tag{1.137}$$

Note that, not coincidentally, the phase shifts in the Cheon–Shigehara model are identical to the phase shifts in the Lieb–Liniger model. This is the basic building block for the Bose–Fermi duality between the two many-body problems.

We are now in position to set the zero-range boundary conditions for the many-fermion wave functions. For each pair of fermions $(i, j)$, we try to generalize condition (1.133) to

$$\lim_{\varepsilon \to 0^+} \frac{d}{d\varepsilon} \int_{x_j - \varepsilon}^{x_j + \varepsilon} dx_i \left[\frac{\partial}{\partial x_i}\psi(x_1,...,x_N) - \frac{\partial}{\partial x_j}\psi(x_1,...,x_N)\right] \overset{?}{=}$$
$$-\frac{1}{a_1}\left[\lim_{x_i \to x_j^+} \psi(x_1,...,x_N) - \lim_{x_i \to x_j^-} \psi(x_1,...,x_N)\right]. \tag{1.138}$$

The relation in equation (1.138) defines the model without the need to refer to a specific interaction. However, and this is important, condition (1.138) for the pair

$(i, j)$ is to be understood for $|x_l - x_j| > 0$ $(l \neq i, j)$, strictly. Otherwise, three-body effects come into play. We can see this with the simplest three-body example with $a_1 \rightarrow \infty$. In that case, the zero-energy state (in the infinite size limit) is given by

$$\psi(x_1, x_2, x_3) = \text{sgn}(x_1 - x_2)\text{sgn}(x_1 - x_3)\text{sgn}(x_2 - x_3). \tag{1.139}$$

The integrand in equation (1.138) takes the form, for pair $(1, 2)$

$$\frac{\partial \psi}{\partial x_1} - \frac{\partial \psi}{\partial x_2} = 4\delta(x_{12}) - 2[\delta(x_{13}) + \delta(x_{23})]. \tag{1.140}$$

Upon integration over $x_1$, for a fixed $\varepsilon > 0$, the integral in equation (1.138) is independent of $\varepsilon$ if $x_3 \notin [x_2 - \varepsilon, x_2 + \varepsilon]$. Therefore, $|x_3 - x_2| > 0$ strictly so that there exists $\eta > 0$ such that $|x_3 - x_2| > \eta$ and, taking $\varepsilon < \eta$ and then the derivative, condition (1.138) is satisfied for this pair. However, if we enforce this, the condition cannot be satisfied for pair $(2, 3)$ unless we include a further condition on three-body coalescence! Hence, we must abandon condition (1.138) in its pure form.

Let us then try something else, again. Take the wave function in equation (1.140), and apply the non-interacting part of the Hamiltonian on it. We have

$$\partial_{x_1}\psi = 2[\delta(x_1 - x_2)\text{sgn}(x_1 - x_3)\text{sgn}(x_2 - x_3) + \text{sgn}(x_1 - x_2)\delta(x_1 - x_3)\text{sgn}(x_2 - x_3)],$$
$$\partial_{x_1}^2\psi = 2[\delta'(x_1 - x_2)\text{sgn}(x_1 - x_3)\text{sgn}(x_2 - x_3) + \text{sgn}(x_1 - x_2)\delta'(x_1 - x_3)\text{sgn}(x_2 - x_3)] \tag{1.141}$$
$$+ 8\delta(x_1 - x_2)\delta(x_1 - x_3)\text{sgn}(x_2 - x_3).$$

Well, well, a three-body term has shown up. The (singular) interaction potential must therefore contain a three-body counterterm. The generalization to $N$ particles is now trivial, for the eigenstate with

$$\psi(x_1,\ldots,x_N) = \prod_{i<j=1}^{N} \text{sgn}(x_i - x_j). \tag{1.142}$$

The interaction has the formal expression

$$V = \frac{2\hbar^2}{m}\sum_{i<j=1}^{N} \delta'(x_i - x_j)\text{sgn}(x_i - x_j) + \frac{4\hbar^2}{m}\sum_{i<j<k=1}^{N} \delta(x_i - x_j)\delta(x_j - x_k). \tag{1.143}$$

Note how strange the above equation looks: it contains undefined products of distributions, in particular Dirac deltas times signum functions. Moreover, it contains a three-body interaction that is purely contact. These interactions are so singular, that they are essentially unusable unless we work out a way to define them properly, as we will do in the next chapter. However, they do tell us that three-body interactions are not negligible in the continuum model, and we should deal with them carefully.

Our next attempt, which will prove free of subtleties, is discretization and subsequent pass to the continuum limit. Let us begin with two particles, with discrete Laplacian including only nearest neighbours, and lattice spacing $d$. Each coordinate can take on values $x_i \in \mathbb{Z}d$. The kinetic energy reads (see volume 1)

$$(H_0\psi)(x_1, x_2) = -J[\psi(x_1 + d, x_2) + \psi(x_1 - d, x_2) + \psi(x_1, x_2 + d) + \psi(x_1, x_2 - d)] \\ + 4J\psi(x_1, x_2). \tag{1.144}$$

Let us work out the interaction, assuming it is local, if the eigenfunction is

$$\psi(x_1, x_2) = \operatorname{sgn}(x_1 - x_2), \tag{1.145}$$

with $\operatorname{sgn}(0) \equiv 0$. Since the space is discrete, defining the sign at zero as zero is meaningful, and includes no issues. For $|x_1 - x_2| > d$, equation (1.144) vanishes for this state. For $x_1 = x_2 \pm d$, we have $(H_0\psi)(x_1, x_2) = 2J\psi(x_1, x_2)$. Therefore, the interaction potential must be

$$V = -2J(\delta_{x_1 + d, x_2} + \delta_{x_1, x_2 + d}), \tag{1.146}$$

where $\delta_{x,y}$ is a Kronecker delta. Equation (1.146) is just a local, two-body, nearest-neighbour interaction. The zero-energy solution is in fact the ground state, corresponding to infinite $a_1$. The non-relativistic continuum limit is obtained by setting $J = \hbar^2/2md^2$ and letting $d \to 0$. Doing the same for three particles, with wave function $\psi = \prod_{i<j=1}^3 \operatorname{sgn}(x_i - x_j)$, we see that no three-body interactions need to be included, and the state remains a good eigenstate, with a well-defined continuum limit. We shall not solve the whole model exactly here—we will do so in the following section—but proceed to extract the short-distance universality of the continuum model from the continuum limit of its discrete version.

Firstly, we will allow a finite scattering length $a_1$ for zero pair total momentum by changing the interaction strength in equation (1.146) from $V_0 = -2J$ to (see volume 1)

$$V_0 = -\frac{2J}{1 - d/a_1} = -\frac{\hbar^2}{md^2}\frac{1}{1 - d/a_1}. \tag{1.147}$$

Then, we write the zero-energy two-body state in the relative coordinate as

$$\phi_2(x_{ij}) = \operatorname{sgn}(x_{ij})[1 - |x_{ij}|/a_1]. \tag{1.148}$$

For $x_{ij} \to cd$, with $d \to 0$ and $c$ any *finite* integer, the many-fermion wave function factorizes as usual

$$\psi(x_1,...,x_N) \to \phi_2(x_{ij})A_2^{(N)}(X_{ij}; \{x_l\}_{l\neq i,j}). \tag{1.149}$$

The first thing we can do is try and derive the adiabatic theorem. Since the Hamiltonian on the lattice is indeed Hermitian, the Hellmann–Feynman theorem will work in its usual version. Hence, we have

$$\frac{\partial E}{\partial(-1/a_1)} = \left\langle \frac{\partial V}{\partial(-1/a_1)} \right\rangle = \frac{2Jd}{(1-d/a_1)^2} \sum_{i<j} \langle \delta_{x_i,x_j+d} + \delta_{x_i,x_j-d} \rangle$$

$$= N(N-1)\frac{2Jd}{(1-d/a_1)^2}\frac{\langle\psi|\delta_{x_i,x_j+d}|\psi\rangle}{\langle\psi|\psi\rangle} \qquad (1.150)$$

$$\to \frac{\hbar^2 C}{m}, \quad d \to 0,$$

Above, we have defined the contact $C$ via

$$C \equiv \lim_{d\to 0}\left[d\sum_x g_2(x, x+d)\right] = \lim_{d\to 0}\left[d\sum_x g_2(x, x-d)\right], \qquad (1.151)$$

which is the discretized version of the continuum contact, equation (1.122). Note that the adiabatic theorem in equation (1.150) is identical to the adiabatic theorem for the Lieb–Liniger model, equation (1.123), which once more points at duality between the two models. The other two relations that are identical to the Lieb–Liniger model are the pressure relation, equation (1.129), and the high-momentum asymptotics of the structure factor, equation (1.126). This is due to the identical short-distance behaviour of $|\phi_2|^2$ in both models. We encourage the reader to prove these facts in the problem section.

However, the momentum distribution for fermions is quite different. This fact is trivial in the sense that the Lieb–Liniger model's momentum distribution for $a_1 = 0^-$ has a tail $\rho(k) \propto C/a_1^2 k^4$, with the contact $C \propto a^2$, while free fermions (corresponding to $a_1 = 0^-$) have a Fermi sea as its momentum distribution. It is very simple to see that, for fermions, the momentum distribution has the following asymptotic behaviour

$$\rho(k) \to \frac{4C}{k^2}, \quad k \to \infty. \qquad (1.152)$$

This is not at odds with the Fermi sea for $a_1 \to 0^-$, since in perturbation theory $C \propto a_1^2$, and the tail disappears at exactly $a_1 = 0^-$.

We must now be careful with the energy theorem. Because of the adiabatic theorem for the Cheon–Shigehara model *in the continuum limit*, equation (1.150), and because the Hamiltonian on the lattice is Hermitian, we have

$$E = \lim_{d\to 0}\int_{-\pi/d}^{\pi/d}\frac{dk}{2\pi}\rho(k)[-2J(\cos(kd)-1)] - \frac{\hbar^2 C}{ma_1}. \qquad (1.153)$$

Unfortunately, we cannot exchange the limit and integral signs above! Since the energy theorem is regularization-dependent, results vary depending on the scheme. For instance, Sekino and Nishida [10] showed that in the continuum model,

$$E = \int\frac{dk}{2\pi}\frac{\hbar^2 k^2}{2m}\left[\rho(k) - \frac{4C}{k^2}\right] + \frac{\hbar^2 C}{ma_1} + \frac{2\hbar^2 C_3}{m}, \qquad (1.154)$$

where $C_3 = \int dx g_3(x, x, x)$ is the three-body contact. Since the discretized Cheon–Shigehara model is exactly solvable, these two equivalent relations for the total energy provide another method to calculate the three-body contact—or the short-range three-body correlations—provided we can estimate the momentum distribution.

### 1.5.3 Yang–Yang model

The last relevant system is simply the spin-$1/2$ fermionic version the Lieb–Liniger model, with first-quantized Hamiltonian given by equation (1.117). In this case, a pair of fermions in a triplet state do not interact, and therefore we can reduce the interaction to between spin-up and spin-down particles. We shall adopt the same conventions as in the three-dimensional case, where particles $i = 1,...,N_\uparrow$ have spin-up, and particles $i = N_\uparrow,...,N$ have spin-down. The technology needed here is identical to that used for the Lieb–Liniger model, and the whole list of relations can be found in reference [11].

## Problems

1. Consider a two-body interaction potential for identical particles in three dimensions that behaves, asymptotically ($r \to \infty$), as

$$V_\infty(r) = \frac{\hbar^2}{m}\left[\frac{g_6}{r^6} + \frac{g_8}{r^8}\right].$$

Show that the zero-energy $s$-wave state in the relative radial coordinate behaves as

$$a(r) = a + \frac{a_1}{r^3} + \frac{a_2}{r^4} + \frac{a_3}{r^5} + O(r^{-6}),$$

and calculate $a_i$ ($i = 1, 2, 3$) as functions of $g_6$, $g_8$ and the scattering length $a$.
2. Calculate the leading-order asymptotic correction to the finite-energy $s$-wave phase shifts in three dimensions for oscillating interactions with power-law tails, of the form

$$V_\infty(r) = \frac{\hbar^2 g_\nu}{r^\nu} \cos(qr).$$

Which is the smallest power $\nu$ that allows for short-range scattering in this case?
3. Given the following non-self-adjoint Hamiltonian

$$H(\lambda) = H_0 + V(\lambda), \tag{1.155}$$

where $\lambda$ is a parameter:
   (a) Show the generalized Hellmann–Feynman theorem for any eigenvalue $E(\lambda)$ associated with its corresponding right eigenstate $\psi_\lambda$,

$$\frac{\partial E(\lambda)}{\partial \lambda} = \left\langle \frac{\partial V(\lambda)}{\partial \lambda} \right\rangle + \langle \psi_\lambda | V(\lambda) \partial_\lambda \psi_\lambda \rangle - \langle V(\lambda) \psi_\lambda | \partial_\lambda \psi_\lambda \rangle.$$

    (b) Prove the standard Hellmann–Feynman theorem for a self-adjoint Hamiltonian as a corollary.

4. Consider a three-dimensional spin-$1/2$ Fermi gas with LO EFT $s$-wave interactions. Obtain the corresponding pressure relation for an imbalanced system with $N_\uparrow \neq N_\downarrow$ in the thermodynamic limit, that is, with

$$\lim_{N,N_\uparrow \to \infty} \frac{N_\uparrow}{N - N_\uparrow} = c \in \mathbb{Q}.$$

(Hint: the Fermi momenta for each spin component are different, $k_{F\uparrow} \neq k_{F\downarrow}$, which introduces another parameter, namely $\alpha \equiv k_{F\uparrow}/k_{F\downarrow}$.)

5. Derive the generalized virial theorem for the one-dimensional Lieb–Liniger model.

6. Show that Lieb–Liniger bosons and Cheon–Shigehara fermions share the following universal relations:
    (a) Pressure relation.
    (b) Pair correlation function at short distances.
    (c) Static structure factor at large momenta.

7. Obtain all of Tan's relations for the Yang–Yang model, and find the qualitative similarities and quantitative differences between these and those for the Lieb–Liniger model, if $N_\uparrow = N_\downarrow = N/2$.

# References

[1] Zeller S *et al* 2016 Imaging the He$_2$ quantum halo state using a free electron laser *Proc. Natl Acad. Sci.* **113** 14651

[2] Aziz R A, Nain V P S, Carley J S, Taylor W L and McConville G T 1979 An accurate intermolecular potential for helium *J. Chem. Phys.* **70** 4330

[3] Svensson E C, Sears V F, Woods A D B and Martel P 1980 Neutron-diffraction study of the static structure factor and pair correlations in liquid $^4$He *Phys. Rev.* B **21** 3638

[4] Bazak B, Valiente M and Barnea N 2020 Universal short-range correlations in bosonic helium clusters *Phys. Rev.* A **101** 010501

[5] Tan S 2008 Energetics of a strongly correlated Fermi gas *Ann. Phys., NY* **323** 2952

[6] Tan S 2008 Large momentum part of a strongly correlated Fermi gas *Ann. Phys., NY* **323** 2971

[7] Tan S 2008 Generalized virial theorem and pressure relation for a strongly correlated Fermi gas *Ann. Phys., NY* **323** 2987

[8] Kuhnle E D, Hu H, Liu X J, Dyke P, Mark M, Drummond P D, Hannaford P and Vale C J 2010 Universal behavior of pair correlations in a strongly interacting Fermi gas *Phys. Rev. Lett.* **105** 070402

[9] Braaten E 2012 *Universal Relations for Fermions with Large Scattering Length* (Berlin: Springer) p 193

[10] Sekino Y and Nishida Y 2021 Field-theoretical aspects of one-dimensional Bose and Fermi gases with contact interactions *Phys. Rev.* A **103** 043307

[11] Barth M and Zwerger W 2011 Tan relations in one dimension *Ann. Phys., NY* **326** 2544

**IOP** Publishing

Strongly Interacting Quantum Systems, Volume 2
Many-body physics
**Manuel Valiente and Nikolaj Thomas Zinner**

# Chapter 2

# Bose–Fermi mapping and strongly interacting one-dimensional systems

Many realistic many-body systems, ranging from electrons in solids to trapped ultracold atoms, can be described as being essentially one-dimensional. This is particularly clean with ultracold atomic ensembles, for which highly anisotropic optical lattices split the system, physically, into a collection of tight tubes. There, the atoms behave almost as if they were one-dimensional. In volume 1, we saw how scattering properties are modified as transversal confinement is varied, and how purely one-dimensional systems collide. Here, we consider one-dimensional, many-body systems of bosons or fermions focusing, in particular, on what is known as Bose–Fermi duality. That is, the equivalence between certain systems of bosons and fermions in one spatial dimension. We also consider trapped, multicomponent, strongly interacting systems, which can be mapped onto spin chains, with this problem being highly related to the short-distance universality described in chapter 1, as well as a self-contained introduction to Luttinger liquid theory.

## 2.1 Bose–Fermi duality for hard core interactions

In 1960, M D Girardeau published an extremely influential paper entitled *Relationship between Systems of Impenetrable Bosons and Fermions in One Dimension* [1]. In essence, he proved that many-body, non-relativistic systems of spinless bosons and spin-polarized fermions in one dimension, with identical Hamiltonians, are in one-to-one correspondence (under some extra assumptions that we will work out in brief), whenever the Hamiltonian features two-body interactions that have a short-distance hard core. That is, whenever the two-body interaction satisfies $\lim_{x \to 0} V(x) = \infty$. This does not preclude the inclusion of three- or higher-body interactions. This Bose–Fermi duality may appear innocuous at first.

doi:10.1088/978-0-7503-3091-6ch2
2-1

However, the methods we know to treat bosons and fermions are very different and work in very different regimes. Hence, Girardeau's insights allow, in many instances, switching from a bosonic to a fermionic representation and compute many quantities of interest using weak-coupling theory, even if the original theory may be strongly coupled, and vice versa.

Let us work out, as usual, the two-body problem, imposing only the short-distance hard core condition. Assume a two-body fermionic eigenstate of the Hamiltonian $H$ with eigenenergy $E$, in vacuum, is given by $\psi_F(x_1, x_2)$. Obviously, if the interaction has a singularity at the origin but is otherwise smooth (no delta functions or other sneaky distributions) and is local, the wave function satisfies $\psi_F(x, x) = 0$. Recall, from chapter 1, that the main issue with signum distributions is their behaviour at $x = 0$. However, these pose no problem when applied on functions that vanish there. Hence, the following function is perfectly well defined (and vanishes) for $x_1 = x_2$,

$$\psi_B(x_1, x_2) = \text{sgn}(x_1 - x_2)\psi_F(x_1, x_2). \tag{2.1}$$

If there are no other singularities in the potential except at $x_1 = x_2$, and the potential is continuous and differentiable otherwise (for instance, a power-law at short distances, $V(x) \sim 1/x^{2\nu}$, with $\nu$ an integer, as $x \to 0$), then $\psi_F$ is continuous and twice differentiable at $x = 0$. Then, the wave function in equation (2.1) is a *bosonic* solution to the same Schrödinger equation as $\psi_F$, with the same eigenenergy.

The statements in the previous paragraph remain true with more singular hard core potentials. For example, take a two-body hard-rod interaction with diameter $a > 0$, that is,

$$V(x) = V_0\theta(a - |x|), \quad V_0 \to +\infty, \tag{2.2}$$

with $\theta(x)$ the Heaviside step function (not to be confused with the scattering phase shifts). This potential is equivalent to the boundary condition $\psi(x_1, x_2) = 0$ if $|x_1 - x_2| \leqslant a$ for the solutions to the Schrödinger equation. Continuity of the scattering states immediately imply

$$\psi_B(x_1, x_2) = \sin(k|x| - ka)\theta(|x| - a), \tag{2.3}$$

$$\psi_F(x_1, x_2) = \text{sgn}(x)\sin(k|x| - a)\theta(|x| - a). \tag{2.4}$$

That is, the Bose–Fermi mapping is still valid. And it remains valid even if the interaction potential had a hard-rod condition plus another smooth interaction potential. An important example, the point hard core limit ($a \to 0$), is particularly relevant, turning $\psi_F$ into a free fermionic state, and $\psi_B$ into its bosonic dual. Here, the interaction potential can be modelled as

$$V(x) = g\delta(x), \quad g \to \infty, \tag{2.5}$$

and its associated many-body problem is typically called the Tonks–Girardeau limit of the Lieb–Liniger model (although, in all fairness, Girardeau is to be given full credit for this!).

Let us now study the many-body problem. To show the Bose–Fermi duality, we follow Girardeau's original proof [1]. We consider a system of $N$ spinless fermions with a hard core condition ($a \geqslant 0$)

$$\psi_F(x_1, x_2, ..., x_N) = 0, \quad |x_i - x_j| \leqslant a \quad (i \neq j = 1, ..., N), \tag{2.6}$$

and satisfying the Schrödinger equation

$$-\frac{\hbar^2}{2m} \sum_{i=1}^{N} \frac{\partial^2 \psi_F}{\partial x_i^2} + V(x_1, ..., x_N)\psi_F = E\psi_F, \tag{2.7}$$

where $V$ is a regular, local, but otherwise general interaction. The fermionic eigenstate $\psi_F$ is then a continuous function vanishing for $|x_i - x_j| \leqslant a$. Now define a bosonic wave function $\psi_B$ as

$$\psi_B(x_1, x_2, ..., x_N) = \psi_F(x_1, x_2, ..., x_N) \prod_{i<j=1}^{N} \operatorname{sgn}(x_i - x_j). \tag{2.8}$$

Since $\psi_F$ is continuous everywhere, so is $\psi_B$. Moreover, the Schrödinger equation (2.7) is clearly satisfied by the bosonic wave function $\psi_B$ in every order sector (defined by permutations of $x_1 < x_2 ... < x_N$), where the product of signum functions does not change sign. Moreover, if $N$ is odd, and $\psi_F$ satisfies periodic boundary conditions, so does $\psi_B$. Also, in the space of functions that vanish whenever $x_i = x_j$ ($i \neq j = 1, ..., N$), the square of the product of signum distributions is unity. Therefore, the spectra of the bosonic and fermionic problems are identical, and the transformation is norm-preserving: the dynamics of bosons and fermions are in one-to-one correspondence, and therefore duality holds.

We now solve the Tonks–Girardeau gas, for which equation (2.7) has $V \equiv 0$, and the eigenstates satisfy the boundary condition (2.6) with $a = 0$. We take periodic boundary conditions, and an odd particle number $N$. Since, without any need for further boundary conditions $\psi_F$ satisfies equation (2.6), $\psi_F$ is simply a free fermion $N$-body state, given by a Slater determinant. According to the Bose–Fermi mapping, equation (2.8), the bosonic state $\psi_B$ is a Slater determinant times Girardeau's totally antisymmetric function. In every order sector, with non-overlapping positions, $\psi_B$ obeys the (non-interacting) Schrödinger equation. Because this problem is equivalent to a delta-interacting Bose gas with interaction strength $g \to \infty$, we are certainly running into trouble if we try to solve the problem directly for $g \to \infty$ using distributional derivatives. To remedy this, we can choose a number of regularization schemes: (i) use the contact boundary conditions for $1/g = 0$; (ii) place the system on a lattice with a hard core condition $\psi(x_1, x_2, ..., x, ..., x, ..., x_N) = 0$ and take the continuum limit (equivalent to taking $V_0 = 0$ in equation (1.147)); (iii) place the system on a lattice with interaction strength in equation (1.147), $a_1 = -\alpha d$, with $\alpha > 0$, and then take the continuum limit; (iv) consider a system of hard rods with $a > 0$ and carefully take the limit $a \to 0$. Method (iii) is doable (the system is integrable) but overkill. Method (i) is the most straightforward, since the boundary conditions are $\psi$ is continuous, and $\psi = 0$ if $x_i = x_j$ for any pair $i \neq j$. Therefore, $\psi = \psi_F \prod_{i<j} \operatorname{sgn}(x_i - x_j)$, where $\psi_F$ is a Slater

determinant of plane waves. Method (ii) is equally simple, since it is straightforward to check that $\psi = \psi_F \prod_{i<j} \text{sgn}(x_i - x_j)$, where $\text{sgn}(0) = 0$ on the lattice is an eigenstate, and the continuum limit gives exactly the same as with method (i), except on a set of zero (Lebesgue) measures. Method (iv), given a fermionic eigenstate $\psi_F$, was already proven, and is also straightforward. In this case, it is actually the limit that may be interesting, so we follow this.

To do this, take the following fermionic eigenstate

$$\psi_F(x_1, ...,x_N) \equiv \prod_{i<j=1}^{N} \theta(|x_{ij}| - a)\phi_F(x_1, ...,x_N). \tag{2.9}$$

Above, $\phi_F$ is an extension of $\psi_F$ to the inside of the hard cores, which we take to be continuous, twice differentiable everywhere except perhaps when $x_i = x_j$, where it is allowed to be even discontinuous. Applying the kinetic energy operator on $\psi_F$, and working out the resulting expression, we see that if $\phi_F$ has these properties and vanishes when $x_i = x_j$ for all $i \neq j$, the state (2.9) is an eigenstate of the problem. We then take the limit $a \to 0$, which leaves as the only possibility the free fermionic eigenstates.

## 2.2 Bethe Ansatz solution of the Lieb–Liniger model

Before proceeding with further duality relations, we should take a moment to solve the one-dimensional many-boson problem with Dirac delta interactions exactly, using the Bethe ansatz. This is not required for the duality relations, but it proves very convenient, for obvious reasons. We shall follow, for the most part, the original calculation of Lieb and Liniger [2].

Take $N$ bosons in a ring (periodic boundary conditions) with length $L$, and Hamiltonian

$$H = -\frac{\hbar^2}{2m}\sum_{i=1}^{N}\frac{\partial^2}{\partial x_i^2} + g \sum_{i<j=1}^{N} \delta(x_i - x_j). \tag{2.10}$$

We first consider repulsive interactions with $g > 0$. The Dirac delta interaction is equivalent to the following set of boundary conditions on the eigenfunctions of the Hamiltonian $\psi$,

$$\left(\frac{\partial}{\partial x_i} - \frac{\partial}{\partial x_j}\right)\psi|_{x_i = x_j + 0^+} - \left(\frac{\partial}{\partial x_i} - \frac{\partial}{\partial x_j}\right)\psi|_{x_i = x_j - 0^+} = \frac{2mg}{\hbar^2}\psi|_{x_i = x_j}. \tag{2.11}$$

As is quite usual with Bethe solvable models, we consider a particular ordering sector, namely

$$\Gamma_1 = \{(x_1, ...,x_N): 0 \leqslant x_1 \leqslant x_2 \leqslant ... \leqslant x_N \leqslant L\}. \tag{2.12}$$

Since we have spinless bosons, we only need to know the eigenstates in $\Gamma_1$, and we can obtain it in any other ordering sector by (bosonic) symmetry. In the (topological) interior of $\Gamma_1$, $\Gamma_1^o$, the eigenstates satisfy the free Schrödinger equation, that is

$$-\frac{\hbar^2}{2m}\sum_{i=1}^{N}\frac{\partial^2\psi}{\partial x_i^2} = E\psi, \quad (x_1, \ldots, x_N) \in \Gamma_1^o, \tag{2.13}$$

and they satisfy the boundary condition

$$\left(\frac{\partial}{\partial x_{i+1}} - \frac{\partial}{\partial x_i}\right)\psi|_{x_{i+1}=x_i} = \frac{mg}{\hbar^2}\psi|_{x_{i+1}=x_i}, \quad (x_1, \ldots, x_N) \in \Gamma_1. \tag{2.14}$$

To impose the periodic boundary conditions on the ring, we must take the coordinates outside of the region $\Gamma_1$. For any ordering $(x_1, \ldots, x_N)$, the wave function must satisfy

$$\psi(x_1, \ldots, x_i + L, \ldots, x_N) = \psi(x_1, \ldots, x_i, \ldots, x_N), \tag{2.15}$$

and have continuous derivative. If $(0, \ldots, x_N) \in \Gamma_1$, then $(L, \ldots, x_N) \notin \Gamma_1$. However, due to bosonic symmetry, $(x_2, \ldots, x_N, L)$, which is in $\Gamma_1$ is equivalent to $(0, \ldots, x_i, \ldots, x_N)$ and, therefore, the periodic boundary conditions in $\Gamma_1$ read

$$\psi(0, x_2, \ldots, x_N) = \psi(x_2, \ldots, x_N, L), \tag{2.16}$$

$$\partial_x\psi(x, x_2, \ldots, x_N)|_{x=0} = \partial_x\psi(x_2, \ldots, x_N, x)|_{x=L}. \tag{2.17}$$

In the ordering sector $\Gamma_1$, we consider the so-called Bethe ansatz. We choose a set of $N$ quantities, $\{k_i\}_{i=1}^{N}$ with dimensions of inverse length (momentum), that we shall call rapidities or asymptotic momenta, all different from each other. The Bethe ansatz wave function in $\Gamma_1$ has the form

$$\psi(x_1, \ldots, x_N) = \sum_{P} a(P)\exp\left(i\sum_{j=1}^{N}k_{P_j}x_j\right). \tag{2.18}$$

The sum above runs over all possible permutations of *rapidities*, while the coordinates remain in $\Gamma_1$. The coefficients of the expansion, $a(P)$, are the main quantities that we need to calculate, given $\{k_i\}_{i=1}^{N}$ and, as we shall see, they are related very simply to the scattering phase shifts of the theory. If equation (2.18) proves to be an eigenstate (when extended to the whole space of coordinates), then obviously the eigenenergy must be given by

$$E = \frac{\hbar^2}{2m}\sum_{i=1}^{N}k_i^2. \tag{2.19}$$

Since total momentum $K$ is conserved, then it must also hold that $K = \sum_{i=1}^{N}k_i$ is the total momentum of the system. Now, we figure out the relation between $a(P)$ and $a(P')$ for different permutations. Let us consider the identity, $P_1 = I$, which leaves the ordered set $\{k_i\}_{i=1}^{N}$ as is, and a permutation $P_2$ that exchanges the order of $k_2$ and $k_1$. If the Bethe ansatz is indeed a solution, then it must satisfy, according to equation (2.14),

$$i(k_2 - k_1)[a(P_1) - a(P_2)] = \frac{mg}{\hbar^2}[a(P_1) + a(P_2)], \qquad (2.20)$$

which implies

$$\frac{a(P_1)}{a(P_2)} = -\frac{mg/\hbar^2 + i(k_2 - k_1)}{mg/\hbar^2 - i(k_2 - k_1)}. \qquad (2.21)$$

Since the above expression has unit modulus, we can write

$$a(P_1)/a(P_2) \equiv -\exp(-i\tilde{\theta}(k_2 - k_1)), \qquad (2.22)$$

and obtain

$$\tan\left(\frac{\tilde{\theta}(s)}{2}\right) = -\frac{\hbar^2 s}{mg}. \qquad (2.23)$$

The tilded notation for the 'phase shift' $\tilde{\theta}(s)$ is purposeful, since the two-body, even-wave phase shift in the Lieb–Liniger model, $\theta(k) \equiv \theta(k_2 - k_1)/2$ is given by $\tan\theta(k) = 2\hbar^2 k/mg$. Hence, we identify

$$\theta(k) = \frac{\tilde{\theta}(k_2 - k_1)}{2} = \frac{\tilde{\theta}(2k)}{2}. \qquad (2.24)$$

We may arrive at all different permutations by transposing pairs of rapidities in succession. The result is that for each transposition, the new coefficient $a(P)$ gets an extra factor $-\exp(-i\tilde{\theta}_{jk})$, where $j$ and $k$ are the transposed rapidity indices. Applying the periodic boundary conditions, equation (2.16), we extract the particular allowed values of the rapidities via

$$-(-1)^N e^{-ik_j L} = \prod_{l=1}^{N} e^{-i\tilde{\theta}(k_j - k_l)}. \qquad (2.25)$$

Let us analyze the (exponential form of) the Bethe ansatz equations, equation (2.25), in the particularly simple (by now!) Girardeau's limit, $1/g = 0$. In this limit, all the phase shifts vanish, and since $k_i \neq k_j$ for $i \neq j$ (otherwise the wave function would vanish!), we have, for odd $N$,

$$k_j = \frac{2\pi n_j}{L}, \quad n_j \in \mathbb{Z}, \qquad (2.26)$$

The ground state corresponds, obviously, to the following sequence of rapidities

$$\{Lk_j/2\pi\}_{j=1}^{N} = \{-(N-1)/2, -(N-3)/2, \ldots, -1, 0, 1, \ldots, (N-3)/2, (N-1)/2\}. \quad (2.27)$$

That is, the rapidities are simply evenly distributed from $-k_F$ to $+k_F$, as should be expected from Bose–Fermi duality. Keeping now an odd number of particles, and transforming the Bethe ansatz equations to a more convenient form, we have their standard form, namely

$$k_j = \frac{2\pi n_j}{L} + \frac{1}{L} \sum_{1=l(\neq i)}^{N} \tilde{\theta}(k_j - k_l)$$
$$= \frac{2\pi n_j}{L} + \frac{2}{L} \sum_{1=l(\neq i)} \theta(k_{jl}), \tag{2.28}$$

where, on the second line, we have identified $\tilde{\theta}$ with the phase shifts of the model and where $k_{jl} = (k_j - k_l)/2$ are the (actual) relative momenta. To work out the ground state in the thermodynamic limit, we first assume—which is actually true—that the ground state corresponds to equally spaced Bethe numbers (the set of $n_j$'s), between $-k_F$ and $k_F$, as happens for hard core Girardeau's bosons. Then, we use some basic probability theory. We know the probability density, as $L \to \infty$ with $N/L = \rho = k_F/\pi$ fixed, for the set of momenta $k_j^{(0)} \equiv 2\pi n_j/L$, which is given by

$$\Pi(k^{(0)}) = \frac{\Theta(k_F - |k^{(0)}|)}{2k_F}, \tag{2.29}$$

where $\Theta(\cdot)$ is the Heaviside step function. Since the rapidities, or Bethe asymptotic momenta, $k_j$, are functions of $k_j^{(0)}$, in the thermodynamic limit we have $k = k(k^{(0)})$. The probability density for the rapidities is then given by

$$\tilde{\Pi}(k) = \left| \frac{dk^{(0)}}{dk} \right| \Pi(k^{(0)}(k)), \tag{2.30}$$

or, inverting the relation

$$\Pi(k^{(0)}) = \left| \frac{dk}{dk^{(0)}} \right| \tilde{\Pi}(k(k^{(0)})). \tag{2.31}$$

Hence, we can find $q > 0$, in general different from $k_F$, such that $\Pi(k) = 0$ if $|k| > q$. The set of Bethe equations for finite $N$, equation (2.28), becomes an integral equation in the thermodynamic limit, since we can make the replacement

$$\sum_{l=1}^{N} f(k_l) = N \int dk \tilde{\Pi}(k) f(k), \tag{2.32}$$

and the Bethe equations become

$$k = k^{(0)} + 2\rho \int d\kappa \tilde{\Pi}(\kappa) \theta((k - \kappa)/2). \tag{2.33}$$

Differentiating the Bethe equations above with respect to $k$, and using $\rho = k_F/\pi$ we have

$$1 = 2\pi \rho \tilde{\Pi}(k) + \rho \int_{-q}^{q} d\kappa \tilde{\Pi}(\kappa) \theta'((k - \kappa)/2). \tag{2.34}$$

Above, $\tilde{\Pi}(k) = 0$ for $|k| > q$, as we already mentioned. The value of $q$ is to be fixed from the normalization condition

$$\int_{-q}^{q} d\kappa \tilde{\Pi}(\kappa) = 1. \tag{2.35}$$

The Bethe integral equation, equation (2.34), must be solved numerically. However, we may consider the strongly interacting limit and obtain the first few orders in perturbation theory from the Bethe integral equation (2.34). To do this, we begin with the zero-th order term, corresponding to $1/g = \infty$. In this case, $\theta'(k) = 0$ and therefore $\tilde{\Pi}(k) = 1/2\pi\rho$, and $q = k_F$. To leading order in $1/g$, $\theta'(k) = 2\hbar^2 k/g$, and the integral equation becomes

$$1 = 2\pi\rho\tilde{\Pi}(k) - \frac{2\hbar^2\rho}{mg} \int_{-q}^{q} d\kappa \tilde{\Pi}(\kappa) + O(g^{-2}), \tag{2.36}$$

which is solved with $k_F/q = 1 + 2\hbar^2\rho/mg$ and $\tilde{\Pi}(k) = 1/2q$. If we recall that the one-dimensional scattering length is related to $g$ as $g = -2\hbar^2/ma$ (see equation (1.118)), these first order relations simplify to

$$\tilde{\Pi}(k) = \frac{1 - \rho a}{2k_F}, \tag{2.37}$$

$$\frac{q}{k_F} = \frac{1}{1 - \rho a}. \tag{2.38}$$

As one might have expected, the leading order correction to the rapidity distribution gives the exact second-order correction to the ground state energy. Using $E = \sum_{l=1}^{N} \hbar^2 k_l^2/2m \rightarrow N \int dk \tilde{\Pi}(k)\hbar^2 k^2/2m$, we obtain, for the first order energy $E^{(1)}$,

$$E^{(1)} = N\frac{\hbar^2 k_F^2}{6m(1 - \rho a)^2} = \frac{E^{(0)}}{(1 - \rho a)^2}, \tag{2.39}$$

where $E^{(0)}$ is the ground state energy of Girardeau's bosons or, equivalently, non-interacting fermions. To this order, quite interestingly, the ground state energy coincides with the ground state energy for hard core bosons with diameter $a$. However, we have derived these relations for *repulsive* bosons, which means that $a < 0$.

The Bethe ansatz also allows for an exact computation of the thermodynamic properties of the Lieb–Liniger model (or any other Bethe ansatz-solvable model, for that matter). The standard method to do this is due to Yang and Yang, and is called Yang–Yang thermodynamics, or the thermodynamic Bethe ansatz. The arguments to obtain these thermodynamic properties are entropic in nature, and involve somewhat complicated statistical methods (see [3]). However, we may replace these arguments and methods with other far simpler, yet rigorous arguments [4]. These, obviously, also involve entropy, leading to the same occupation probability as in Yang–Yang thermodynamics [3]. Now let us define a real variable with dimensions of momentum (or inverse length), and call it $k^{(0)}$. Given $T$, $L$ and $\mu$, let us suppose that $k^{(0)} = 2\pi n/L$, with $n$ being integers, is distributed according to the following probability density

$$\Pi_\beta(k^{(0)}) = A \frac{1}{e^{\beta(\varepsilon(k^{(0)})-\mu)} + 1}. \tag{2.40}$$

Above, $A$ is a normalization constant, and $\varepsilon(k^{(0)})$ is a symmetric, positive, smooth function of $k^{(0)}$, which also depends implicitly on the temperature through $\beta = 1/k_B T$. Equation (2.40) is the maximum entropy distribution for $k^{(0)}$, if all $k^{(0)}$'s have to be different from each other (therefore the Fermi distribution) and, since $k^{(0)}$ is discretized on a regular grid, we can take the thermodynamic (continuum) limit, $L \to \infty$, so that $A = 1/2\pi\rho(\mu)$, where $\rho = \rho(\mu)$ is the density, which depends on the chemical potential. We now define $\varepsilon(k^{(0)}) = \hbar^2 k^2/2m$, where $k$ and $k^{(0)}$ are related via the Bethe ansatz equations (in their discrete form). This means that, in the continuum limit, we have, for the (internal) energy $E$ of the system

$$E = N \int dk \tilde{\Pi}_\beta(k) \frac{\hbar^2 k^2}{2m} = N \int dk^{(0)} \frac{1}{2\pi\rho(\mu)} \frac{\varepsilon(k^{(0)})}{e^{\beta(\varepsilon(k^{(0)})-\mu)} + 1}, \tag{2.41}$$

and $\rho(\mu)$ is fixed by normalization, that is $\int dk \tilde{\Pi}_\beta(k) = 1$. Putting everything together, and proceeding as for the ground state, we obtain the following integral equation

$$G(k) = 1 + \int \frac{d\kappa}{2\pi} \frac{\theta'(k - \kappa)G(\kappa)}{\exp[\beta(\hbar^2\kappa^2/2m - \mu)] + 1}, \tag{2.42}$$

where $G$ is defined via

$$\tilde{\Pi}_\beta(k) = \frac{G(k)/2\pi\rho(\mu)}{\exp[\beta(\hbar^2 k^2/2m - \mu)] + 1} \tag{2.43}$$

The thermodynamic Bethe equation, together with the normalization condition, must be solved numerically. But once more, we can do some perturbation theory in $1/g$, near Girardeau's point. If $1/g = 0$, then $G(k) = 1$, and the thermodynamics in Girardeau's hard core limit is, of course, identical to that of a non-interacting Fermi gas. To leading order in $1/g$, equation (2.42) is solved by

$$G(k) = \frac{1}{1 + I(\beta, \mu)a}, \tag{2.44}$$

where

$$I(\beta, \mu) = \int \frac{d\kappa}{2\pi} \frac{1}{\exp[\beta(\hbar^2\kappa^2/2m - \mu)] + 1}. \tag{2.45}$$

Imposing normalization on $\tilde{\Pi}_\beta$, we obtain

$$I(\beta, \mu) = \frac{\rho(\mu)}{1 + \rho(\mu)a}, \tag{2.46}$$

and, therefore

$$\tilde{\Pi}_\beta(k) = \frac{1 - \rho(\mu)a}{2\pi\rho(\mu)} \frac{1}{\exp[\beta(\hbar^2 k^2/2m - \mu)] + 1}, \tag{2.47}$$

that is, the distribution is equivalent to that of a free Fermi gas with rescaled density $\tilde{\rho} = \rho/(1 - \rho a)$.

Now that we have completely solved for the energetics and thermodynamics of the Lieb–Liniger model, and obtained the leading order corrections to $O(1/g)$, we can make some predictions about the contact which, in turn, gives us information about short-distance and large-momentum asymptotics of correlation functions. Using the adiabatic theorem, equation (1.123), we obtain for the contact in the ground state

$$\frac{\hbar^2 C}{m} = 2aE^{(0)}\frac{\rho a}{(1 - \rho a)^3}, \tag{2.48}$$

which is exact to $O((\rho a)^2)$. We can also extract the pressure (to this order) in the ground state using the pressure relation, equation (1.129), obtaining

$$P = \frac{2\mathcal{E}^{(0)}}{(1 - \rho a)^3}, \tag{2.49}$$

where $\mathcal{E}^{(0)} = E^{(0)}/L$ is the non-interacting fermionic ground state energy density. Correlation functions, beyond the short-distance limit, can be calculated using the Bethe ansatz wave functions. However, these are rather complicated calculations that go beyond the scope of this book, and we refer the reader to the research literature on this topic [5].

## 2.3 Attractive Lieb–Liniger model. McGuire's solution

The attractive Lieb–Liniger model is also an exactly solvable problem. The ground state—called McGuire's solution—in vacuum, is a deeply bound state representing a bright soliton. Although McGuire's original solution, reference [6], is very instructive, we present here a much easier route to obtaining the ground state.

Consider the following operators $A_i$ ($i = 1, ...,N$),

$$A_i = \frac{\partial}{\partial x_i} + \lambda \sum_{1=j(\neq i)}^{N} \text{sgn}(x_i - x_j), \tag{2.50}$$

and the following SuSy Hamiltonian (see volume 1),

$$H_{\text{SuSy}} = \frac{\hbar^2}{2m}\sum_{i=1}^{N} A_i^\dagger A_i. \tag{2.51}$$

Since $A_i^\dagger A_i$ is a positive operator, the ground state energy of $H_{\text{SuSy}}$ is bounded from below as $E_0 \geqslant 0$. Hence, if we can find $\psi$ such that $A_i\psi = 0$ for each $i = 1, ...,N$, and $\psi$ is normalizable, then it is the many-body ground state. It is trivially verified that

$$\psi(x_1, \ldots, x_N) = \prod_{i<j=1}^{N} e^{-\lambda|x_i - x_j|} \qquad (2.52)$$

is annihilated by all $A_i$'s. Therefore, it is the ground-state energy of $H_{\text{SuSy}}$ and its energy is zero if $\lambda > 0$. We now expand $H_{\text{SuSy}}$ and find that

$$H_{\text{SuSy}} = \sum_{i=1}^{N} \frac{p_i^2}{2m} - \frac{\hbar^2\lambda}{m}\sum_{i\neq j}\delta(x_i - x_j) + C \qquad (2.53)$$

Therefore, the attractive Lieb–Liniger Hamiltonian is given by $H_{\text{SuSy}}$ up to an offset $C$, identifying the coupling constant $g = -2\hbar^2\lambda/m$. From this constant (see problem 4), we obtain the ground state energy of the Lieb–Liniger model as $E_0 = -C$.

## 2.4 Spinless Bose–Fermi duality without hard cores

In chapter 1, we looked as carefully as possible at the issues involving a low-energy theory for one-dimensional interacting fermions. The resulting model, called the Cheon–Shigehara model, requires some heavy regularization and renormalization, and we saw that the only easy way of achieving this was by means of lattice discretization and the subsequent pass to the continuum limit. In that case, it appears quite obvious that there must be a one-to-one Bose–Fermi mapping in the continuum limit between the Lieb–Liniger and Cheon–Shigehara models. This will be the first proof we work out below. The continuum version of the model, without placing the particles on a lattice, however, has its own interest, and we would be missing out on a lot of nice and heavy machinery if we did not go through the proof of Bose–Fermi duality directly in the continuum. And we shall do this in two different ways: (i) standard regularization–renormalization and (ii) generalized function algebras. The first method, while quite standard *a priori*, requires the inclusion of three-body forces. The second method, much less standard, gives essentially a free proof of duality. However, it is very formal and not computationally friendly.

### 2.4.1 Proof I. Bosons and fermions on a lattice

We begin by restating the lattice Hamiltonian that lead to a well-defined continuum limit for fermions in the form of the Cheon–Shigehara model. This is given, in second quantization, by

$$H = -J\sum_{j}(c_j^\dagger c_{j+1} + c_{j+1}^\dagger c_j - 2c_j^\dagger c_j) + V_0\sum_{j}n_j n_{j+1}. \qquad (2.54)$$

Above, $J = \hbar^2/2md^2$ is the tunnelling rate, $c_j$ $(c_j^\dagger)$ is a fermionic annihilation (creation) operator at lattice site $j$ (with $j$ integer, and the position of the particles $x = jd$), while

$V_0 = -2J/(1 - d/a)$ is the nearest-neighbour interaction strength (see equation (1.147)).

What is usually done in the lattice model is to apply the so-called Jordan–Wigner transformation, which maps fermions onto hard core bosons. The reader might be disappointed, but can be reassured by the fact that, in the continuum limit, the points at which particles meet form a set of vanishing (Lebesgue) measures, so the hard cores are not relevant, if the interaction is properly taken care of. Instead of the Jordan–Wigner transformation, which is a mapping in second quantization, we are going to use first quantization and Girardeau's mapping on the lattice. The action of the Hamiltonian on a first-quantized wave function is given by

$$(H\psi)(x_1, ...,x_N) = -J\sum_{i=1}^{N}\sum_{\mu=\pm 1} \psi(x_1, ...,x_i + \mu d, ...,x_N) + 2JN\psi(x_1, ...,x_N)$$

$$+ V_0 \sum_{i<j=1}^{N} [\delta_{x_i,x_j + d} + \delta_{x_i,x_j - d}]\psi(x_1, ...,x_N). \tag{2.55}$$

Since we are dealing with spinless fermions, $\psi$ must be antisymmetric under the exchange of any two-particle coordinates, which implies that $\psi(x_1, ...,x, ...,x, ...,x_N) = 0$. The only issue for the Bose–Fermi mapping could occur when two or more particles neighbour each other. However, if we set

$$\psi_B(x_1, ...,x_N) = \psi(x_1, ...,x_N) \prod_{i<j=1}^{N} \text{sgn}(x_i - x_j), \tag{2.56}$$

with $\text{sgn}(0) = 0$, we see, immediately, that for an odd number of particles and periodic boundary conditions, the action of $H$ on the bosonic wave function $\psi_B$ is identical to its action on the fermionic one $\psi$, equation (2.55). Therefore, since the local transformation $\prod_{i<j} \text{sgn}(x_i - x_j)$ is unitary in the subspace of totally symmetric (bosons) or totally antisymmetric (fermions) $N$-body wave functions with a hard core condition $\psi(x_1, ...,x, ...,x, ...,x_N) = 0$. This finishes the proof of the Bose–Fermi mapping on the lattice.

To prove that the resulting eigenstates in the continuum limit correspond to those of the Lieb–Liniger model for bosons, and therefore the Cheon–Shigehara model for bosons, we may solve the stationary Schrödinger equation using the Bethe ansatz, and prove that in the continuum limit the bosonic Bethe wave functions become those of the Lieb–Liniger model except on a set with vanishing Lebesgue measure, in the zero-filling limit (corresponding to finite density in the continuum). Let us then solve the hard core Bose gas problem on the lattice using the Bethe ansatz. We proceed just as in the Lieb–Liniger model. Take a region $\Gamma_1$, defined as

$$\Gamma_1 = \{(x_1, ...,x_N) \in \mathbb{Z}d: x_1 < x_2 ... <x_N \}. \tag{2.57}$$

We do not need to include points with $x_i = x_j$ since the wave function vanishes there. In $\Gamma_1$, we write

$$\psi_B(x_1, ...,x_N) = \sum_P a(P) e^{i\sum_{j=1}^{N} k_{P_j} x_j}, \qquad (2.58)$$

which is identical to what we had in the Lieb–Liniger model, with the same notation. But now, instead of differentiating the wave function to satisfy the boundary condition, we take any two particles, say $x_1$ and $x_2$, and put them one lattice site apart, that is, $x_2 = x_1 + d$. At that point in $\Gamma_1$, the action of the Hamiltonian on the wave function, using the hard core condition, is given by

$$
\begin{aligned}
(H\psi_B)(x_1, x_1 + d, x_3, \dots, x_N) &= -J\big[\psi_B(x_1 - d, x_1 + d, \dots, x_N) \\
&\quad + \psi_B(x_1, x_1 + 2d, \dots, x_N)\big] + \cdots \\
&\quad + 4J\psi_B(x_1, x_1 + d, \dots, x_N) \\
&\quad + V_0\psi_B(x_1, x_1 + d, \dots, x_N) + \cdots .
\end{aligned}
\qquad (2.59)
$$

The dots '$\cdots$' above correspond to the action of the Hamiltonian on particles $j > 2$. Assume, first, that $x_{i+1} - x_i > d$ for all $i \neq 1$, and that $x_N - x_1 \neq (L_s - 1)d$, with $L_s$ the number of lattice sites. Hence the contributions marked with '$\cdots$' are trivial, in the sense that they give

$$... = H_0^{(N-2)}\psi_B = -2J\sum_P a(P)\left[\exp\left(i\sum_{l=1}^{N} k_{P_l} x_l\right)\sum_{j=3}^{N}[\cos(k_{P_j} d) - 1]\right] \qquad (2.60)$$

Clearly, for $\psi_B$ to satisfy the Schrödinger equation at these points of $\Gamma_1$, we must have

$$
\begin{aligned}
&- J\left[\psi_B(x_1 - d, x_1 + d, ...,x_N) + \psi_B(x_1, x_1 + 2d, ...,x_N)\right] \\
&+ 4J\psi_B(x_1, x_1 + d, ...,x_N) + V_0\psi_B(x_1, x_1 + d, ...,x_N) \\
&= -2J\sum_P a(P)\left[\exp\left(i\sum_{l=1}^{N} k_{P_l} x_l\right)\sum_{j=1}^{2}[\cos(k_{P_j} d) - 1]\right],
\end{aligned}
\qquad (2.61)
$$

which is trivially verified. Since the phase shifts in the lattice model have Lieb–Liniger's phase shifts as continuum limit, we easily conclude that the former is equivalent to the latter in this limit.

### 2.4.2 Proof II. Non-abelian algebra

We proceed now to use an algebra of generalized functions to prove Bose–Fermi duality. While this can be done for more general effective field theory (EFT) interactions, including but not limited to three-body interactions, we will only present the proof for two-body low-energy (LO) EFT interactions, that is, Lieb–Liniger to Cheon–Shigehara duality. Generalized function algebras are used to regularize and renormalize problems in a very direct, yet formal manner. The strange-looking products are nothing but the consequence of *one* particular choice of regularization, but the final, renormalized results, are independent of this particular choice.

The algebra of generalized functions we discuss here is due to Shirokov—Shirokov's algebra, from now on. The formal math behind Shirokov's algebra is rather complicated, and we only provide here some naïve justification. The interested reader is referred to Shirokov's original work in reference [7].

We define a product (*) of generalized functions such that

$$\text{sgn}(x) * \text{sgn}(x) = 1, \quad \forall\, x \in \mathbb{R}. \tag{2.62}$$

If we impose Leibniz's rule, then

$$0 = \frac{d}{dx}[\text{sgn}(x) * \text{sgn}(x)] = 2\delta(x) * \text{sgn}(x) + 2\text{sgn}(x) * \delta(x), \tag{2.63}$$

that is, the delta and signum generalized functions *anticommute*, or

$$\{\text{sgn}(x), \delta(x)\} = 0. \tag{2.64}$$

We now differentiate $\delta(x) * \text{sgn}(x)$, obtaining

$$\frac{d}{dx}[\delta(x) * \text{sgn}(x)] = \delta'(x) * \text{sgn}(x) + 2\delta(x) * \delta(x)$$
$$= -\frac{d}{dx}[\text{sgn}(x) * \delta(x)] = -2\delta(x) * \delta(x) - \text{sgn}(x) * \delta'(x). \tag{2.65}$$

Hence, the anticommutator

$$\{\delta'(x), \text{sgn}(x)\} = -4\delta(x) * \delta(x). \tag{2.66}$$

We take the anticommutator above, and * -multiply it with the signum, obtaining

$$\delta'(x) + \text{sgn}(x) * \delta'(x) * \text{sgn}(x) + 4\delta(x) * \delta(x) = 0 \tag{2.67}$$

Now, the first two generalized functions above have negative parity, the third one has positive parity. Hence, $\delta(x) * \delta(x)$ vanishes identically, and we have the two identities

$$\delta(x) * \delta(x) = 0, \tag{2.68}$$

$$\{\delta'(x), \text{sgn}(x)\} = 0. \tag{2.69}$$

In fact, it is now easy to prove that

$$\{\delta^{(m)}(x), \text{sgn}(x)\} = 0, \quad \forall\, m \in \mathbb{N}. \tag{2.70}$$

We already have all we need to prove Bose–Fermi duality. The innocent looking regularization of the product of two signum functions in equation (2.62) accomplishes something very useful: Girardeau's Bose–Fermi mapping, which is a local unitary transformation for hard core particles, becomes unitary for *arbitrary* particles! The only price to pay is that all products become Shirokov algebra's * -products.

Take the Lieb–Liniger Hamiltonian,

$$H = \sum_{i=1}^{N} \frac{p_i^2}{2m} + g \sum_{i<j=1}^{N} \delta(x_i - x_j). \tag{2.71}$$

Let $\psi_B$ be a bosonic eigenstate of $H$. Then, $\psi_F = \prod_{i<j} \mathrm{sgn}(x_i - x_j)\psi_B$ is an eigenstate of $H_F$, given by

$$
\begin{aligned}
H_F = &-\frac{\hbar^2}{2m} \sum_{i=1}^{N} \prod_{l<k=1}^{N} \mathrm{sgn}(x_l - x_k) \frac{\partial^2}{\partial x_i^2} \left[ \left( \prod_{l'<k'=1}^{N} \mathrm{sgn}(x_{l'} - x_{k'}) \right) \cdot \right] \\
&+ g \sum_{i<j=1}^{N} \left( \prod_{l<k=1}^{N} \mathrm{sgn}(x_l - x_k) \right) \delta(x_i - x_j) \left( \prod_{l'<k'=1}^{N} \mathrm{sgn}(x_{l'} - x_{k'}) \right),
\end{aligned}
\tag{2.72}
$$

where all singular products of distributions are to be taken as $*$-products. The alert reader may immediately point out that if we take two particles, for simplicity, then the even-wave Dirac delta interaction becomes, for fermions

$$g\,\mathrm{sgn}(x_1 - x_2)\delta(x_1 - x_2)\mathrm{sgn}(x_1 - x_2) = -g\delta(x_1 - x_2), \tag{2.73}$$

because of the anticommutation relation between signum and Dirac delta. It then may appear as if fermions do not feel this even-wave interaction. However, take a two-body fermionic eigenstate, with energy $E$, $\psi_F(x_1, x_2) = \mathrm{sgn}(x_1 - x_2)\psi_B(x_1, x_2)$, and enact equation (2.73) on it. We obtain

$$-g\delta(x_1 - x_2)\mathrm{sgn}(x_1 - x_2)\psi_B(x_1, x_2) = g\,\mathrm{sgn}(x_1 - x_2)\delta(x_1 - x_2)\psi_B(x_1, x_2). \tag{2.74}$$

Now apply the kinetic energy, in the relative coordinate $x = x_1 - x_2$, on $\psi_F(x_1, x_2) = \exp(iKX)\mathrm{sgn}(x)\psi_B(x)$, and we obtain

$$-\frac{\hbar^2}{m}\mathrm{sgn}(x)\frac{\partial^2}{\partial x^2}\psi_B, \tag{2.75}$$

so that

$$H_F\psi_F(x_1, x_2) = \mathrm{sgn}(x_1 - x_2)E\psi_B(x_1, x_2) = E\psi_F(x_1, x_2). \tag{2.76}$$

Hence, indeed, the even-wave, Dirac delta interaction does not have a trivial action on fermionic states! This is solely due to Shirokov's algebraic product. If we generalize this to the many-body problem, and we take $\psi_F = [\prod_{i<j} \mathrm{sgn}(x_i - x_j)]\psi_B$, with $\psi_B$ a bosonic eigenstate of the Lieb–Liniger model with energy $E$, we see immediately that $H\psi_F = E\psi_F$.

The above analysis proves that there is such unitary transformation between bosons and fermions. Then, obviously, the unitarily transformed Hamiltonian has the fermionic dual states as eigenstates. But this is not very useful, and does not connect with previous analyses. To correct this, let us rewrite $H_F$ in equation (2.72) as $H_F = H_0 + W + V$, with $H_0$ the non-interacting Hamiltonian. We obtain

$$V = -g \sum_{i<j=1}^{N} \delta(x_{ij}), \tag{2.77}$$

$$W = -\frac{2\hbar^2}{m} \sum_{i<j=1}^{N} [\delta'(x_{ij})S(x_{ij})] - \frac{4\hbar^2}{m} \sum_{i<j<k=1}^{N} \delta(x_{ij})\delta(x_{ik}). \tag{2.78}$$

The three-body term is not necessary in this formalism. However, note that a three-body counter term must be included—and removed appropriately in the end—when regularizing and renormalizing the $N$-body problem for $N > 2$. This was done with standard renormalization in reference [8], and with either Danilov–Pricoupenko filtering or with a counter term in chapter 7 of volume 1.

### 2.4.3 Proof III. Continuum regularization and renormalization

We already have clues as to what the fermionic two-body interaction should look like in order to achieve Bose–Fermi duality. We may do this by regularizing the two-body interaction from Shirokov's algebra, using LO fermionic EFT or using some other regularization scheme. In order to connect with the regularization in the three-fermion sector, we will use a space representation of the effective interaction as a regularization.

In the relative coordinate $x = x_1 - x_2$, let us regularize a contact interaction as

$$V(x) = V_0 \Theta(b - |x|), \tag{2.79}$$

where $V_0$ is the interaction strength, and $b$ the diameter of the interaction. To obtain a Dirac delta interaction with scattering length $a$, in the zero-range limit, $b \to 0^+$, we set the interaction to

$$V(x) = -\frac{\hbar^2}{mba} \Theta(b - |x|) \to g\delta(x), \quad b \to 0^+. \tag{2.80}$$

To see what the regularized interaction looks like in the context of LO EFT, we obtain the momentum representation of the interaction, given by

$$\langle k'|V|k\rangle = g\frac{\sin(qb)}{qb}, \tag{2.81}$$

where $q = k - k'$. The odd-wave interaction, $V_o(k', k)$ is obtained as

$$V_o(k', k) = \frac{1}{2}[\langle k'|V|k\rangle - \langle k'|V|-k\rangle], \tag{2.82}$$

which reads, explicitly

$$V_o(k', k) = g\frac{k' \sin(kb)\cos(k'b) - k \sin(k'b)\cos(kb)}{(k^2 - k'^2)b}. \tag{2.83}$$

Expanding $V_o(k', k)$ at low-momentum transfers $k'$ and $k$, to LO, we obtain the desired effective interaction

$$V_o^{\mathrm{LO}}(k', k) = \frac{gb}{2}k'k \equiv g_1 k'k, \tag{2.84}$$

which, as expected, concides with the usual LO EFT interaction in the odd-wave channel that we saw in volume 1. In equation (2.84), we have defined the bare coupling constant $g_1 \equiv g_1(\Lambda)$, since $b$ is a spatial parameter, and the interaction (2.84) is momentum represented, and will introduce a sharp cutoff $\Lambda \propto 1/b$, but the proportionality constant has to be calculated consistently.

Let us work out the two-fermion problem now to fix $g_1(\Lambda)$ as a function of the scattering length. The Lippmann–Schwinger equation for the $T$-matrix has the form

$$\langle k'|T(z)|k\rangle = g_1 k'k + g_1 \int_{-\Lambda}^{\Lambda} \frac{dq}{2\pi} k'q \frac{\langle q|T(z)|k\rangle}{z - \hbar^2 q^2/m}. \tag{2.85}$$

Clearly, the $T$-matrix takes the form $\langle k'|T(z)|k\rangle = t(z)k'k$, and equation (2.85) reduces to

$$t(z) = g_1 + g_1 \int_{-\Lambda}^{\Lambda} \frac{dq}{2\pi} q^2 \frac{t(z)}{z - \hbar^2 q^2/m}, \tag{2.86}$$

which is solved by

$$t(z) = \frac{1}{1/g_1 - I(z, \Lambda)}, \tag{2.87}$$

where

$$I(z, \Lambda) = \int_{-\Lambda}^{\Lambda} \frac{dq}{2\pi} \frac{q^2}{z - \hbar^2 q^2/m}. \tag{2.88}$$

The result of the above integral, on-shell ($z = \hbar^2 k^2/m + i0^+$) is listed in appendix A in volume 1, and is given by

$$I(\hbar^2 k^2/m + i0^+, \Lambda) = -\frac{m\Lambda}{\pi\hbar^2} - i\frac{mk}{2\hbar^2} + O(\Lambda^{-1}). \tag{2.89}$$

We use the following renormalization condition, $t(0) = g_{1,R}$, with $g_{1,R}$ the renormalized coupling constant. Then, the bare coupling constant takes the form, as $\Lambda \to \infty$,

$$\frac{1}{g_1} = \frac{1}{g_{1,R}} - \frac{m\Lambda}{\pi\hbar^2}. \tag{2.90}$$

We match the one-dimensional, odd-wave scattering length $a$ by using equations (5.22) and (5.24) in volume 1, with

$$g_{1,R} = \frac{2\hbar^2 a}{m} = -4\left(\frac{\hbar^2}{m}\right)^2 \frac{1}{g}, \qquad (2.91)$$

which gives the relation between the fermionic renormalized coupling constant, $g_{1,R}$ and the bosonic coupling constant $g$.

The issues at the three-particle level are completely removed by using the regularized interaction in equation (2.80) with finite $b$. The interaction strength, however, must be adjusted accordingly. For fermion–fermion scattering states, we solve the two-body problem at zero energy. Using this regularization, the entire issue appears to rely on fine tuning, as we require the interaction strength, for small $b$, to go as $\sim \alpha b^{-(2+\varepsilon)}$, with $\varepsilon > 0$. The limit, as $b \to 0$, of the scattering length vanishes regardless of $\alpha$. However, the scattering length as a function of $b$, depending on the choice of $\varepsilon$ diverges finitely or infinitely many times as we approach $b = 0$, which allows for fine tuning to an arbitrary value of our choice (see problem section). It is best, in a simulation, to choose small but finite $b$.

## 2.5 Richardson–Gaudin models

Since recently, experimental capabilities with ultracold atomic systems have been enhanced to the point of being able to prepare, manipulate and perform measurements on very small systems. In particular, the issue of how superconducting pairing is built up as the particle number increasing from a few-body to a many-body scale has been investigated experimentally [9]. One way to deal with fermion–fermion pairing in small systems, leading to the usual Bardeen–Cooper–Schrieffer (BCS) theory as $N \to \infty$, consists of considering a model due to Richardson [10], and further refined mathematically by Gaudin. The set of models, all of them integrable, that can be built, fall under the category of Richardson–Gaudin models. These can be formulated in arbitrary dimensions but admit a one-dimensional magnetic representation, due to Gaudin (the Gaudin magnet). Therefore, we study these models here. We follow two very thorough references in the topic, namely the review by Dukelsky, Pittel and Sierra [10] (more physically oriented), and the excellent PhD thesis by Claeys [11] (more mathematically oriented).

### 2.5.1 Gaudin magnet

Let us consider a set of $L$ Hermitian operator charges, $Q_i$, that will allow us to write down an integrable spin-1/2 chain. Defining the $SU(2)$ Pauli matrices at site $j$ ($=1, 2, ..., L$) $\sigma_j^\alpha$, $\alpha = x, y, z$, the most general quadratic operator can be written as

$$Q_i = \sum_{j(\neq i)=1}^{N} \sum_{\alpha=x,y,z} W_{ij}^\alpha \sigma_i^\alpha \sigma_j^\alpha, \qquad (2.92)$$

where $W_{ij}^\alpha$ are real coefficients. A necessary condition for the system we would like to construct to be integrable is the commutativity of charges, that is,

$[Q_i, Q_j] = 0 \; \forall \; i \neq j$. A lengthy but straightforward calculation results in the condition ($\alpha \neq \beta \neq \gamma \neq \alpha$ and $i \neq j \neq k \neq i$ below)

$$W_{ij}^\alpha W_{jk}^\gamma + W_{ji}^\beta W_{ik}^\gamma - W_{ik}^\alpha W_{jk}^\beta = 0. \tag{2.93}$$

A particular solution that solves equation (2.93) consists of imposing $W_{ij}^\alpha = -W_{ji}^\alpha$ (antisymmetry, just as in the construction of Lax matrix entries, see volume 1). Moreover, for $S^z$-conserving systems, that require $[S^z, Q_i] = 0$ ($S^z = \sum_i \sigma_i^z/2$), we further obtain

$$W_{ij}^x = W_{ij}^y \equiv X_{ij}, \tag{2.94}$$

$$W_{ij}^z \equiv Y_{ij}. \tag{2.95}$$

With the above relation, together with condition (2.93), one admissible solution that is of interest to BCS pairing is $X_{ij} = Y_{ij}$, with

$$X_{ij} = \frac{1}{\varepsilon_i - \varepsilon_j}, \tag{2.96}$$

where $\varepsilon_i$ ($i = 1, 2, ...,L$) are real parameters. We can now immediately write down a Hamiltonian as a linear combination of the $L$ charges $Q_i$ as

$$H = 2\sum_{i=1}^{L}\xi_i H_i. \tag{2.97}$$

By construction, $H$ commutes with all charges and all charges commute with each other. We can rewrite $H$, up to a constant offset, in a more convenient manner as (see problem 2)

$$H = \sum_{i \neq j}\frac{\xi_i - \xi_j}{\varepsilon_i - \varepsilon_j}\vec{\sigma_i} \cdot \vec{\sigma_j}, \tag{2.98}$$

where $\vec{\sigma_i} = (\sigma_i^x, \sigma_i^y, \sigma_i^z)$ is the vector of spin-1/2 Pauli matrices. To connect the Gaudin magnet to the pairing problem, we may add an innocuous term to each of the charges (since we have included $S^z$-conservation), simply as

$$Q_i \to \tilde{Q}_i + S_i^z, \tag{2.99}$$

$$H \to \tilde{H} + \sum_i \xi_i S_i^z. \tag{2.100}$$

We shall slightly abuse the notation and write $H$ and $Q_i$ when we mean $\tilde{H}$ and $\tilde{Q}_i$ from now on.

### 2.5.2 Superconducting Richardson model

Let us consider a system of fermions in the second quantization. It is described by $L$ energy degrees of freedom, which we call energy levels. Each energy level is labelled

$l = 1, 2, ..., L$, and is $\Omega_l$-fold degenerate. Each state within the degenerate subspace in level $l$ is labelled $(lm_l)$, and we define the so-called pseudospin $S_l$ via the multiplicity of the state as $\Omega_l \equiv 2S_l + 1$. Within a level, time-reversed partners come in pairs as $(lm)$ and $(l\bar{m})$.

After taking all the considerations above into account, let us define the following operators

$$\tau_l^z = \frac{1}{2}\sum_m c_{lm}^\dagger c_{lm} - \frac{\Omega_l}{4}, \tag{2.101}$$

$$\tau_l^+ = \frac{1}{2}\sum_m c_{lm}^\dagger c_{l\bar{m}}^\dagger, \tag{2.102}$$

$$\tau_l^- = \frac{1}{2}\sum_m c_{l\bar{m}} c_{lm}. \tag{2.103}$$

Above, $c_{lm}^\dagger$ ($c_{lm}$) is a fermionic creation (annihilation) operator in state $(lm)$, that is

$$\{c_{lm}, c_{l'm'}^\dagger\} = \delta_{ll'}\delta_{mm'}. \tag{2.104}$$

The operator $\tau_l^z$ plays the role of number operator in level $l$, while $\tau_l^+$ ($\tau_l^-$) creates (destroys) a paired state in level $l$ with a time-reversed pair. We now calculate all commutators, which show that $\tau_l^\pm$ are ladder operators for the $SU(2)$ algebra, while $\tau_l^z$ is a spin-1/2 Pauli matrix. Therefore, these operators close the $SU(2)$ algebra. We evaluate one of the commutators explicitly, mainly because the notation can be rather confusing. In equation (2.101), the index $m$ in the sum runs through *all* possible values of $m$ within the $l$-level. However, in equations (2.102) and (2.103), the index $m$ runs only through half of the values, for otherwise the time-reversed partners would appear twice in the sum and with opposite sign, and $\tau_l^\pm$ would be identically zero. As a simple track keeping device, we use $m > 0$ in equations (2.102) and (2.103), and all $m \neq 0$ for $\tau_l^z$. Then, we can simply denote the time-reversed partner of $(lm)$ as $(l\bar{m}) \equiv (l, -m)$. The commutator is easily evaluated by observing that

$$[c_{lm}^\dagger c_{lm}, c_{l'm'}^\dagger c_{l'\bar{m}'}^\dagger] = \delta_{ll'}\left(\delta_{mm'} c_{lm}^\dagger c_{l\bar{m}}^\dagger + \delta_{m\bar{m}'} c_{l\bar{m}}^\dagger c_{lm}^\dagger\right). \tag{2.105}$$

Using the above relation, we obtain, explicitly

$$[\tau_l^z, \tau_{l'}^+] = \frac{\delta_{ll'}}{4}\left[\sum_{m>0} c_{lm}^\dagger c_{l\bar{m}}^\dagger + \sum_{m<0} c_{l\bar{m}}^\dagger c_{lm}^\dagger\right] = \frac{\delta_{ll'}}{2}\sum_{m>0} c_{lm}^\dagger c_{l\bar{m}}^\dagger = \delta_{ll'}\tau_l^+. \tag{2.106}$$

Similar calculations lead to the standard relations for a $SU(2)$ algebra, that we summarize as

$$[\tau_l^z, \tau_{l'}^\pm] = \pm\delta_{ll'}\tau_l^\pm, \tag{2.107}$$

$$[\tau_l^+, \tau_{l'}^-] = 2\delta_{ll'}\tau_l^z. \tag{2.108}$$

As promised earlier in the section, we can build a pairing (or Richardson) Hamiltonian as a Gaudin magnet. We simply write down the Gaudin magnet Hamiltonian, equation (2.98) including $\sum_l \xi_l S_l^z$ (see equation (2.100)), in terms of the fermionic field operators, that is

$$H = \sum_{i \neq j} \frac{\xi_i - \xi_j}{\varepsilon_i - \varepsilon_j} \vec{\tau_i} \cdot \vec{\tau_j} + \frac{1}{2}\sum_i \xi_i \tau_i^z. \tag{2.109}$$

Using $\tau_i^\pm = \tau_i^x \pm i\tau_i^y$, we obtain

$$H = \sum_{i \neq j} \frac{\xi_i - \xi_j}{\varepsilon_i - \varepsilon_j}\left(\frac{1}{4}(\tau_i^+ + \tau_i^-)(\tau_j^+ + \tau_j^-) - \frac{1}{4}(\tau_i^+ - \tau_i^-)(\tau_j^+ - \tau_j^-) + \tau_i^z \tau_j^z\right)$$
$$+ \frac{1}{2}\sum_i \xi_i \tau_i^z \tag{2.110}$$
$$= \sum_{i \neq j} \frac{\xi_i - \xi_j}{\varepsilon_i - \varepsilon_j}(\tau_i^+ \tau_j^- + \tau_i^z \tau_j^z) + \frac{1}{2}\sum_i \xi_i \tau_i^z.$$

Inserting the definitions of $\tau_i^\pm$ and $\tau_i^z$ in terms of the fermionic operators, equations (2.101), (2.102) and (2.103), and defining $n_{jm} = c_{jm}^\dagger c_{jm}$, we obtain

$$H = \frac{1}{4}\sum_{i \neq j} \frac{\xi_i - \xi_j}{\varepsilon_i - \varepsilon_j} \sum_{mm'>0}\left(c_{im}^\dagger c_{i\bar{m}}^\dagger c_{j\bar{m}} c_{jm} + (n_{im} + n_{i\bar{m}})(n_{jm'} + n_{j\bar{m}'})\right)$$
$$+ \frac{1}{4}\sum_i \xi_i \sum_{m>0}(n_{im} + n_{i\bar{m}}). \tag{2.111}$$

The Hamiltonian (2.111) starts to look like a pairing Hamiltonian, albeit in an usual form. We need to make some simplifications, and a choice. First, we note that the number of fermions in level $j$ is just $N_j = \sum_{m \neq 0} n_{jm}$, so that the rightmost term in equation (2.111) simplifies to $\sum_i \xi_i N_i/4$, and the second term becomes $N_i N_j$. Next, we make the choice $\xi_i = g\varepsilon_i$, with $g$ a coupling constant, and the Richardson Hamiltonian becomes

$$H = \frac{g}{4}\sum_{i \neq j}\sum_{m,m'>0} c_{im}^\dagger c_{i\bar{m}}^\dagger c_{j\bar{m}'} c_{jm'} + \frac{g}{4}\sum_{i \neq j} N_i N_j + \frac{1}{4}\sum_i \xi_i N_i$$
$$= \frac{g}{4}\sum_{i \neq j}\sum_{m,m'>0} c_{im}^\dagger c_{i\bar{m}}^\dagger c_{j\bar{m}'} c_{jm'} + \sum_i \xi_i N_i + \frac{g}{4}N^2 - \frac{g}{4}\sum_i N_i^2, \tag{2.112}$$

where we have defined $N = \sum_i N_i$. Well, well, isn't that pretty? A nice and simple-looking BCS-type Hamiltonian, called Richardson's model. We now know—or at least have a huge suspicion—that this model is integrable. It turns out it is also exactly solvable. How do we solve it? Clearly, we should turn back to Gaudin's magnet, which we solve next and connect its solution to the specifics of the pairing system. Just to have a cleaner Hamiltonian, we may remove the last two terms in equation (2.112), since they are constants of motion, and redefine the Hamiltonian as

$$H = \frac{g}{4} \sum_{i \neq j} \sum_{m,m'>0} c_{im}^\dagger c_{i\bar{m}}^\dagger c_{j\bar{m}'} c_{jm'} + \sum_{i,m} \xi_i c_{im}^\dagger c_{im}. \tag{2.113}$$

### 2.5.3 Exact solution of the Gaudin magnet and Richardson's model

In section 2.5.1 we showed the existence of $L$ conserved charges for Gaudin's magnet. We have not proved that they are independent, but this can be proved (see [11]). To solve the problem exactly, we try and find simultaneous eigenstates of all charges $Q_i$. Since the Hamiltonian $H$ is a linear combination of charges, then any simultaneous eigenstate of all $Q_i$ ($i = 1, 2, ...,L$) is also an eigenstate of $H$. Hence, we would try to solve

$$Q_i |q_1, q_2, ...,q_L\rangle = q_i |q_1, q_2, ...,q_L\rangle, \quad i = 1, 2, ...,L. \tag{2.114}$$

The eigenvalues $q_i$—the values taken by the conserved charges—will be called rapidities, since the solution will take the form of a Bethe ansatz. Following [10], we write down the solution in the fermionic representation (i.e., for Richardson's pairing model). We define the following unpaired state

$$|\nu\rangle = |\nu_1, ...,\nu_L\rangle, \tag{2.115}$$

with $\nu = \sum_j \nu_j$, such that $A_j |\nu\rangle = 0$ and $N_j |\nu\rangle = \nu_j |\nu\rangle$. It is now easy to see that the eigenstates take the form

$$|\psi\rangle = B_1^\dagger ... B_M^\dagger |\nu\rangle, \tag{2.116}$$

where the operators $B_\alpha^\dagger$ are given by

$$B_\alpha^\dagger = \sum_j \frac{1}{\varepsilon_j - E_\alpha} A_j^\dagger, \tag{2.117}$$

and the so-called Richardson's roots satisfy the following equation

$$1 - 4g \sum_j \frac{\nu_j/2 - \Omega_j/4}{2\varepsilon_j - E_\alpha} + 4g \sum_{\beta(\neq\alpha)} \frac{1}{E_\alpha - E_\beta} = 0. \tag{2.118}$$

Last but not least, the energy associated with the eigenstate is given by

$$E = \sum_j \varepsilon_j \nu_j + \sum_\alpha E_\alpha. \tag{2.119}$$

## 2.6 Luttinger liquids

In higher dimensions ($D > 1$), there is a low-energy theory for fermionic quasiparticles, called Fermi liquid theory, in which excitations are single-particle entities with a finite lifetime [12]. This theory, which is very standard, will not be described in this book. In one spatial dimension, however, Fermi liquid theory breaks down. This is mainly due to non-analyticities, which can already be seen in perturbation theory. Moreover, one-dimensional systems are interesting on their own, and, even more importantly, the distinction between bosons and fermions is rather blurred in this case. Therefore, we attempt to build an effective low-energy theory which should be capable of describing bosons and fermions alike. This is what is called Luttinger liquid theory. Out of the many routes that we can take to develop the theory, we will follow the constructive path. Firstly, we will define the massless Thirring model. Then, we will proceed to explain the *bosonization* technique, and show how generic, non-relativistic gases may be described using the effective model, even though the dispersion relations are not exactly linear anywhere.

### 2.6.1 Massless Thirring model

The massless Thirring model is a simple relativistic quantum field theory (QFT) in $(1 + 1)$ dimensions. It features massless fermions coupled via a contact interaction. We can write down its Hamiltonian density, defined as ($\hbar = 1$ and the speed of light $c = 1$)

$$\mathcal{H} = -i\left(\psi_1^\dagger \partial_x \psi_1 - \psi_2^\dagger \partial_x \psi_2\right) + 2g\psi_1^\dagger \psi_2^\dagger \psi_2 \psi_1, \tag{2.120}$$

where $\psi = (\psi_1, \psi_2)$ is a two-dimensional fermionic spinor, such that $\{\psi_i(x), \psi_j^\dagger(x')\} = \delta_{ij}\delta(x - x')$, and $g$ is a coupling constant. It turns out that the massless Thirring model is exactly solvable. Its solution, as trivial as it may seem at first sight, is plagued with subtleties and must be regularized and renormalized with care. We place the system on a ring with length $L$, so its Hamiltonian $H$ is simply

$$H = \int_0^L dx \mathcal{H}(x). \tag{2.121}$$

In a condensed matter language, the massless Thirring model corresponds to two fermionic fields with linear dispersion relation interacting via a zero-range potential. One of the fermionic components moves rightwards, and the other component moves leftwards. Collisions only occur between a left-moving and a right-moving fermion. In the first quantization, the two-body problem describing elastic collisions between a 1- and a 2-fermion amounts to solving the stationary Schrödinger equation

$$-i[\partial_{x_1}\phi(x_1, x_2) - \partial_{x_2}\phi(x_1, x_2)] + g\delta(x_1 - x_2)\phi(x_1, x_2) = E\phi(x_1, x_2), \tag{2.122}$$

where $\phi$ is the coefficient of the two-body, second-quantized wave function $|\Phi\rangle$,

$$|\Phi\rangle = \int_0^L dx_1 \int_0^L dx_2 \phi(x_1, x_2)\psi_1^\dagger(x_1)\psi_2^\dagger(x)|0\rangle, \tag{2.123}$$

where $|0\rangle$ is the particle vacuum, defined via $\psi_i(x)|0\rangle = 0$ $(i = 1, 2)$ $\forall$ $x \in [0, L]$. The Schrödinger equation is separated into centre-of-mass $(X = (x_1 + x_2)/2)$ and relative $(x = x_1 - x_2)$ coordinates, as usual, so that

$$\phi(x_1, x_2) = \exp(iKX)\phi(x), \tag{2.124}$$

and equation (2.122) becomes

$$-2i\partial_x\phi(x) + g\delta(x)\phi(x) = E\phi(x). \tag{2.125}$$

To connect with the first-quantized version of the problem completely, we write down the full first-quantized wave functions as

$$|\Phi\rangle = \phi(x_1, x_2)|12\rangle - \phi(x_2, x_1)|21\rangle. \tag{2.126}$$

We take the limit $L \to \infty$ to solve for scattering states before attempting the calculation. Let us first assume that instead of a delta interaction, we replace $g\delta(x)$ with a smooth potential $V(x)$. Then

$$\phi(x) = \exp\left[iEx/2 - i\int dx\, V(x)/2\right]. \tag{2.127}$$

That is, the wave function simply picks up a phase as it goes through the potential. We try to use this for the Dirac delta function, for which $\int V(x)dx = g\theta(x)$, where $\theta(x)$ is the Heaviside step function. We would obtain

$$\phi(x) = \exp\left[iEx/2 - ig\theta(x)/2\right]. \tag{2.128}$$

We must now be very careful when operating using the exponential form (2.128). Let us try to differentiate it using the standard chain rule. We would obtain

$$\phi'(x) \overset{?}{=} i\frac{E}{2}\phi(x) - i\frac{g}{2}\delta(x)e^{-ig\theta(x)/2}. \tag{2.129}$$

Now we have a problem. The first term on the right-hand-side of equation (2.129) is fine, but the second term contains the product of two distributions. As such, it is undefined. It may well be $-ig\delta(x)/2$ or $-ig\delta(x)\exp(-ig/2)$ or, in fact, anything we want it to be. Let us try a different path. Since, for all $x \neq 0$, the wave function (2.128) takes the form

$$\phi(x) = e^{iEx/2}[\theta(-x) + \theta(x)e^{-ig/2}], \tag{2.130}$$

we may differentiate the above expression so that

$$\phi'(x) = i\frac{E}{2}\phi(x) + \delta(x)(e^{-ig/2} - 1). \tag{2.131}$$

Now, at the very least, we see that $\phi'$ remains a distribution! Equation (2.130) is indeed the correct way of defining the eigenstate. This is not so satisfying, since how could we choose one form or the other a *priori*? The answer comes from the theory of distributions itself. Let us forget about usual derivatives, and instead work out the weak derivative of the exponential representation of $\phi$ or, to simplify things further, of $\tilde{\phi}(x) = \exp(-ig\theta(x)/2)$. We write, for a smooth function $f$ that vanishes as $x \to \pm\infty$,

$$\langle f | \partial_x \tilde{\phi} \rangle = \equiv - \int_{-\infty}^{\infty} dx \partial_x f^*(x) \tilde{\phi}(x) = - \int_{-\infty}^{\infty} dx \partial_x f^*(x) e^{-ig\theta(x)/2}$$

$$= - \int_{-\infty}^{0} dx \partial_x f^*(x) - \int_{0}^{\infty} dx \partial_x f^*(x) e^{-ig/2} = -f^*(0) + f^*(0)e^{-ig/2} \quad (2.132)$$

$$= f^*(0)(e^{-ig/2} - 1).$$

We see that this is indeed the derivative we were looking for, that is,

$$\tilde{\phi}'(x) = \delta(x)(e^{-ig/2} - 1). \quad (2.133)$$

This means that we identify, as distributions

$$\theta(-x) + e^{-ig/2}\theta(x) = \exp(-ig\theta(x)/2), \quad (2.134)$$

and their derivatives are $\delta(x)(\exp(-ig/2 - 1))$. We still have one problem, though. Although we saw, naïvely, that the wave function is an eigenstate, in reality it is not when written as in equation (2.130). Besides that, the wave function $\phi(x)$ is a distribution and so is $\delta(x)$. Therefore, their product remains undefined. We invoke Shirokov's algebra. Proceeding just as with the signum function, we find that $[\theta(x)]^2 = \theta(x)$ and, differentiating it and applying Leibniz's rule, we obtain

$$\delta(x) = \delta(x)\theta(x) + \theta(x)\delta(x). \quad (2.135)$$

Using that $\theta(x) + \theta(-x) = 1$ (as a distribution or generalized function), we have

$$\delta(x)\theta(-x) = \theta(x)\delta(x). \quad (2.136)$$

Therefore, we obtain

$$\delta(x)[\theta(x) + \theta(-x)] = \delta(x), \quad (2.137)$$

or, equivalently, that every time we encounter $\theta(0)$ we may set it to $1/2$. Now we are ready to try once more to obtain an appropriate eigenstate, by setting

$$\phi(x) = e^{iEx/2}[\theta(-x) + \lambda\theta(x)]. \quad (2.138)$$

Inserting the above wave function into the Schrödinger equation, we find that

$$\lambda = \lambda(g) = e^{-i\xi(g)}, \quad (2.139)$$

$$\xi(g) = 2\arctan(g/4). \quad (2.140)$$

We see now that the problem, upon appropriate use of distributions and regularization–renormalization, yields a well-defined answer. We can check this even further by going to the momentum representation and solving the two-body problem. The $T$-matrix satisfies the Lippmann–Schwinger equation

$$T(z) = V + VG_0(z)T(z), \quad (2.141)$$

which, on-shell ($z = 2k + i0^+$), after identifying $g$ with $g_0$—the bare coupling constant, takes the form

$$T(z) = \cfrac{1}{1/g_0 - \int \cfrac{dq}{2\pi} \cfrac{1}{2k - 2q + i0^+}} = \frac{1}{1/g_0 + i/4}. \qquad (2.142)$$

We now obtain the position representation of the scattering wave function. The scattered wave $\phi_s$ is given by

$$|\phi_s\rangle = G_0(z)T(z)|k\rangle \implies \phi_s(x)$$
$$= \frac{1}{1/g_0 - i/4} \int \frac{dq}{2\pi} \frac{e^{iqx}}{2k - 2q + i0^+} = -\theta(x)\frac{i/2}{1/g_0 + i/4}e^{ikx}, \qquad (2.143)$$

so the scattering wave function reads, in full

$$\phi(x) = e^{ikx}\left[1 - \frac{i/2}{1/g_0 + i/4}\theta(x)\right] = e^{ikx}\left[\theta(-x) + \frac{1 - ig_0/4}{1 + ig_0/4}\theta(x)\right] \qquad (2.144)$$
$$= e^{ikx}[\theta(-x) + e^{-i\xi(g_0)}\theta(x)],$$

with

$$\xi(g_0) = 2\arctan(g_0/4). \qquad (2.145)$$

Hence, the renormalization condition reads $g_0 = g$, and we have arrived at the same result. Note, however, that in position space special care (using Shirokov's algebra) was required. In momentum space, we have assumed that $k$ is not of $O(\Lambda)$, where $\Lambda$ is a hard cutoff. Otherwise, the real part of the inverse of $T(z)$ would not vanish identically. In the Thirring model this is not a minor point: the 'Fermi sea' is filled and therefore there are arbitrarily large-momentum scales involved in the problem.

Now that we have solved the two-body problem, we could try to attempt a Bethe ansatz solution. This, however tempting, is not so trivial, because of the non-trivial scattering at high energies. In fact, we do not know of an appropriate solution to the Bethe ansatz dilemma, unless the constant phase shift is assumed, which is only true at finite energies. We may do this, but keeping well in mind that this is not a rigorous procedure for Thirring's Hamiltonian, as it is instead the solution to a ficticious integrable model with scale invariance unbroken at all scales. If we proceed this way, we can write down the (component-symmetric) Bethe ansatz solution with momenta $(k_1, ..., k_N, q_1, ..., q_N)$ as

$$|\Phi\rangle = \int dx_1 ... dx_N \int ds_1 ... ds_N \prod_{i=1}^{N} e^{ik_ix_i} \prod_{j=1}^{N} e^{iq_jx_j}$$
$$\prod_{i',j'=1}^{N} [\theta(s_{j'} - x_{i'}) + e^{-i\xi(g)}\theta(x_{i'} - s_{j'})] \qquad (2.146)$$
$$\times \prod_{i=1}^{N} \psi_1^\dagger(x_i)^N \prod_{j=1}^{N} \psi_2^\dagger(s_j)|0\rangle.$$

The Bethe equations with periodic boundary conditions take the form

$$k_i = \frac{2\pi n_i}{L} + \frac{2}{L}\sum_j \arctan(g/4), \tag{2.147}$$

$$q_j = -\frac{2\pi m_i}{L} - \frac{2}{L}\sum_i \arctan(g/4). \tag{2.148}$$

Above, $n_i$ and $m_i$ are integers. The ground state (vacuum in QFT) is obtained by selecting a cutoff $\Lambda = 2\pi M/L$, with $M = \tilde{N}/2$ and $\tilde{N}$ the number of particles (this is where this procedure differs from the actual Thirring model), and considering the set of Bethe numbers $n_i = -M + 1, -M + 1, ...,0$ and $m_i = 0, 1, ...,M - 1$. The ground state energy is nothing but

$$E_0(\tilde{N}) = \sum_{i=1}^{M} k_i - \sum_{j=1}^{M} q_j = \frac{\tilde{N}}{L}\arctan(g/4) - \frac{4\pi}{L}\sum_{n=1}^{M} n$$

$$= -\frac{\pi\tilde{N}}{L}\left(\frac{\tilde{N}}{2} - 1\right) + \frac{\tilde{N}}{L}\arctan(g/4). \tag{2.149}$$

Since $\tilde{N}$ is only related to the cutoff $\Lambda = \pi\tilde{N}/L$, we rewrite the vacuum energy density as

$$\frac{E_0(\Lambda)}{L} = -\Lambda\left(\frac{\Lambda}{2\pi} - \frac{1}{L}\right) + \frac{\Lambda}{\pi}\arctan(g/4). \tag{2.150}$$

We can now extract the speed of sound (in QFT the speed of light) by considering a particle–hole excitation at the Fermi point. We remove a particle from $n_i = 0$ and set it at $n_i = 1$. The difference in energy is simply $\omega = 2\pi/L$, which implies that the speed of light is unchanged, $v = 1$, as should be for a Lorentz-invariant system.

### 2.6.2 Bosonization of the Thirring and Tomonaga–Luttinger models

The first thing we can do is write down its Hamiltonian in the momentum representation, namely

$$H = \sum_k k(c_{k1}^\dagger c_{k1} - c_{k2}^\dagger c_{k2}) + \frac{g}{L}\sum_{k,k',q} c_{k+q,\,1}^\dagger c_{k'-q,\,2}^\dagger c_{k',2} c_{k,1}. \tag{2.151}$$

We perform a canonical transformation to particle and hole operators, as

$$\begin{aligned}
c_{k1} &= b_k, \quad k \geqslant 0, \\
c_{k1} &= a_k^\dagger, \quad k < 0, \\
c_{k2} &= b_k, \quad k < 0, \\
c_{k2} &= a_k^\dagger, \quad k \geqslant 0.
\end{aligned} \tag{2.152}$$

Upon applying the transformations above, the non-interacting part of Thirring's Hamiltonian (2.151) becomes

$$H_0 = \sum_k |k|\left(a_k^\dagger a_k + b_k^\dagger b_k\right) + E^{(0)}, \tag{2.153}$$

where $E^{(0)}$ is the (divergent) energy of the non-interacting vacuum

$$E^{(0)} = \sum_{k<0} k - \sum_{k>0} k. \tag{2.154}$$

Since $E^{(0)}$ amounts to a constant energy shift (however infinite), we drop it from here on.

We find now the bosonic operators relevant to bosonization. Define, for $q > 0$,

$$\rho_\sigma(q) \equiv \sum_k c_{k+q,\,\sigma}^\dagger c_{k\sigma}, \tag{2.155}$$

$$\rho_\sigma(-q) \equiv \sum_k c_{k\sigma}^\dagger c_{k+q,\sigma}, \tag{2.156}$$

for $\sigma = 1, 2$. By writing them down in terms of the particle–hole operators, equation (2.152), we find that

$$[\rho_\sigma(-q),\, \rho_{\sigma'}(q')] = \delta_{\sigma,\sigma'}\delta_{q,q'}\frac{qL}{2\pi}. \tag{2.157}$$

With these definitions, the interaction is written as

$$V = \frac{g}{L}\sum_{q>0}[\rho_1(-q)\rho_2(q) + \rho_1(q)\rho_2(-q)] + \frac{g}{L}N_1 N_2, \tag{2.158}$$

where $N_\sigma$ is the number of $\sigma$ fermions. Next, we show that, up to a constant, the non-interacting part of the Hamiltonian can also be written as a quadratic form in terms of $\rho_\sigma(\pm q)$. This is easy to see since

$$[H_0, \rho_1(\pm q)] = \pm q\rho_1(\pm q), \tag{2.159}$$

$$[H_0, \rho_2(\pm q)] = \mp q\rho_2(\pm q). \tag{2.160}$$

Therefore, up to a constant offset, we have

$$H_0 = \frac{2\pi}{L}\sum_{q>0}[\rho_1(q)\rho_1(-q) + \rho_2(-q)\rho_2(q)]. \tag{2.161}$$

We now define bosonic operators $A_1(q)$ and $A_2(-q)$, via

$$A_1(q) = \left(\frac{2\pi}{qL}\right)^{1/2}\rho_1(-q), \tag{2.162}$$

$$A_2(-q) = \left(\frac{2\pi}{qL}\right)^{1/2}\rho_2(q), \tag{2.163}$$

so we can write the Hamiltonian, up to a constant offset, as

$$H = \sum_{q>0} q \left[ A_1^\dagger(q)A_1(q) + A_2^\dagger(-q)A_2(-q) \right.$$

$$\left. + \frac{g}{2\pi}(A_1(q)A_2(-q) + A_1(q)^\dagger A_2(-q)^\dagger) \right]. \tag{2.164}$$

Since the Hamiltonian above is quadratic in bosonic operators, we can diagonalize it using a Bogoliubov transformation. We write

$$\tilde{A}_1(q) = A_1(q)\cosh\delta - A_2^\dagger(-q)\sinh\delta, \tag{2.165}$$

$$\tilde{A}_2(-q) = A_2(-q)\cosh\delta - A_1^\dagger(q)\sinh\delta. \tag{2.166}$$

To leave the problem diagonal, we have to impose

$$K_L \equiv e^{2\delta} = \sqrt{\frac{1 - g/2\pi}{1 + g/2\pi}}, \tag{2.167}$$

where we have called this quantity $K_L$, the Luttinger parameter. In diagonal form, except for irrelevant constants, we obtain

$$H = u \sum_{q>0} \left[ \tilde{A}_1^\dagger(q)\tilde{A}_1(q) + \tilde{A}_2^\dagger(-q)\tilde{A}_2(-q) \right], \tag{2.168}$$

where $u$ is the speed of excitations, given by

$$u = \sqrt{1 - (g/2\pi)^2}. \tag{2.169}$$

This result is both a success and a failure. If we are trying to obtain an effective theory for a system that is *not* relativistic (as we shall do below), then surely equation (2.169) is a success: the speed of excitations must change due to interactions. If we are trying to diagonalize Thirring's model itself, then it is a catastrophic failure, for the speed of excitations—the speed of light—*cannot* change: we must have 'inadvertently' broken Lorentz invariance! Well, in fact, even in the non-relativistic case, further renormalization (we did not need to renormalize anything to get a finite result) is required. The speed of light in the bare Hamiltonian was set to $c = 1$. However, we need Thirring's model to give $u = 1$ once the full solution is obtained, which means that we should renormalize (without infinities!) the bare speed of light, and set it to $c \to \sqrt{1 + (g/2\pi)^2}$. Then, we immediately obtain $u = 1$, as should be. For non-relativistic effective models, this renormalization also needs to be done, in a different way, and also has to do with (Galilean) relativity.

### 2.6.3 Haldane's bosonization

The general usage of Luttinger liquid theory as a low-energy effective field theory for fermions and bosons in one spatial dimension is due to Haldane, who wrote an extremely influential article in 1981 [13]. The physics is very neat, and the resulting low-energy theory is just Luttinger's model. Here, we follow both Haldane's pioneering work as well as Giamarchi's book on one-dimensional 'quantum' physics [14].

We begin by assuming we have a one-dimensional system of particles. The first-quantised, space-represented density operator if the particles are located at fixed positions $\{x_j\}_j$ is given by

$$\rho(x) = \sum_j \delta(x - x_j). \tag{2.170}$$

We assume the particles are displaced by $\Delta_j$ from their classical crystalline ground state configurations, corresponding to $j/\rho_0$, where $\rho_0$ is the mean density. That is, $x_j = j/\rho_0 + \Delta_j$. We define a smooth classical field $\varphi(x)$ such that $\varphi(x_i) = 2\pi j$. That is, $\varphi$ assigns particle labels to individual positions, and is arbitrary (but smooth) everywhere else. We need the properties of Dirac delta functions, in particular

$$\delta(\varphi(x) - 2\pi n) = \sum_i \frac{\delta(x - x_i)}{|\partial_x \varphi(x)|}. \tag{2.171}$$

We must also use the Fourier transform of the periodic delta function,

$$\delta(s - 2\pi n) = \sum_{\ell \in \mathbb{Z}} \frac{e^{i\ell s}}{2\pi}. \tag{2.172}$$

Combining these two properties, we obtain for the density operator

$$\rho(x) = \frac{\partial_x \varphi(x)}{2\pi} \sum_{\ell \in \mathbb{Z}} e^{i\ell \varphi(x)}. \tag{2.173}$$

Since the particle positions are not fixed, it is convenient to define a new operator, $\phi = \pi \rho_0 x - \frac{1}{2}\varphi$, so that the density operator can be rewritten as

$$\rho(x) = \left[\rho_0 - \frac{1}{\pi}\partial_x \phi(x)\right] \sum_{\ell \in \mathbb{Z}} \exp\left[i\ell(2\pi\rho_0 x - 2\phi(x))\right]. \tag{2.174}$$

When taking space-averages, the higher harmonics ($|\ell| \geqslant 1$) in equation (2.174), which are highly oscillating, vanish, and we end up with the following smeared density

$$\rho(x) \sim \rho_0 - \frac{1}{\pi}\partial_x \phi(x). \tag{2.175}$$

We now define the following creation operator

$$\psi_B^\dagger(x) = \sqrt{\rho(x)}\, e^{-i\theta(x)}, \tag{2.176}$$

where $\theta(x)$ is some phase operator yet to be determined, and $\rho(x)$ is, at this time, the smeared density, equation (2.175). The subscript 'B' stands for bosonic, and we will try to construct $\theta(x)$ such that $\psi_B^\dagger$ is a bosonic creation operator, that is,

$[\psi_B(x), \psi_B^\dagger(x')] = \delta(x - x')$. This commutation relation can be satisfied to linear order in density fluctuations with respect to the mean density $\rho_0$. We define $\Pi(x) \equiv -\partial_x\phi(x)/\pi$, and define $\theta(x)$ via

$$[\theta(x), \Pi(x')] = i\delta(x - x'), \tag{2.177}$$

that is, as the conjugate of the fluctuation operator $\Pi(x)$. We then expand the square root of the smeared density to $\mathcal{O}(\Pi)$,

$$\sqrt{\rho(x)} = \sqrt{\rho_0} + \frac{1}{2\sqrt{\rho_0}}\Pi(x) + \mathcal{O}(\Pi^2). \tag{2.178}$$

Dropping higher-order terms in the commutator, we obtain

$$[\psi_B(x), \psi_B^\dagger(x')] = \frac{1}{2}[e^{-i\theta(x)}, \Pi(x')e^{i\theta(x')}] + \frac{1}{2}[e^{-i\theta(x)}\Pi(x), e^{i\theta(x')}]. \tag{2.179}$$

Since $[\theta(x), \Pi(x')] = i\delta(x - x')$, we have $[f(\theta(x)), \Pi(x')] = i\delta(x - x')f'(\theta(x))$ and, using this in equation (2.179), we obtain the desired result, $[\psi_B(x), \psi_B^\dagger(x')] = \delta(x - x')$.

We now reintroduce the higher harmonics into the density operator, equation (2.174), obtaining for the unsmeared bosonic creation operator

$$\psi_B^\dagger(x) = \left[\rho_0 - \frac{1}{\pi}\partial_x\phi(x)\right]^{1/2}\sum_{\ell\in\mathbb{Z}}\exp[i\ell(2\pi\rho_0 x - 2\phi)(x)]e^{-i\theta(x)}. \tag{2.180}$$

So far, we have only dealt with bosons. To connect with the rest of the section, where fermions are being considered (although, as we have seen, for non-relativistic particles bosons and fermions are alike), we define a fermionic creation operator $\psi^\dagger(x)$, as

$$\psi^\dagger(x) \equiv \psi_B^\dagger(x)e^{i\varphi(x)/2}. \tag{2.181}$$

It appears natural that $\psi^\dagger(x)$ is fermionic, sice $\varphi(x_j)/2 = \pi j$, and so $\exp(i\varphi(x_j)/2)$ is a phase that changes sign every time a particle's position is crossed. But to see that it is indeed a fermionic operator, we calculate its anticommutation relation,

$$\{\psi(x), \psi^\dagger(x')\} = e^{-i\varphi(x)/2}\psi_B(x)\psi_B^\dagger(x')e^{i\varphi(x')/2} + \psi_B^\dagger(x')e^{i\varphi(x')/2}e^{-i\varphi(x)/2}\psi_B(x). \tag{2.182}$$

We need the commutator $[\theta(x), \phi(x')/\pi]$. But from equation (2.177), we have that

$$\partial_{x'}[\theta(x), \phi(x')/\pi] = i\delta(x - x') \implies [\theta(x), \phi(x')/\pi] = \frac{1}{2}\mathrm{sgn}(x' - x). \tag{2.183}$$

Therefore, we obtain, after working out all the commutators carefully,

$$\{\psi(x), \psi^\dagger(x')\} = e^{-i\phi(x)}[\psi_B(x), \psi_B(x')]e^{i\phi(x')} = \delta(x - x'). \tag{2.184}$$

Let us now see what the Hamiltonian should look like. If we have a non-relativistic (Galilean) kinetic energy, then obviously for bosons, if we take the leading order contributions (necessary for commutation relations to hold!),

$$\partial_x \psi_B^\dagger \partial_x \psi_B \rightarrow \rho_0 \partial_x(e^{-i\theta(x)})\partial_x(e^{i\theta(x)}) = \rho_0 [\partial_x \theta(x)]^2. \tag{2.185}$$

Of course, the coefficient in front of this part of the Hamiltonian is renormalized and, using dimensional analysis, we have

$$H_0 = \frac{\hbar}{2} \int dx \frac{uK}{\pi}[\partial_x\theta(x)]^2, \tag{2.186}$$

where $K$ is a dimensionless constant and $u$ has dimensions of velocity, and both depend on the microscopic details of the target many-body system. As for the interaction energy, we may use a Dirac-delta interaction for bosons, so, to leading order (again, necessary for commutation relations to hold), we have, up to an irrelevant energy offset,

$$V = \frac{g_0}{2} \int dx [\rho(x)]^2 = \frac{g_0 \rho_0}{2} \int dx \Pi(x) + \frac{g_0}{2} \int dx [\Pi(x)]^2. \tag{2.187}$$

Since $\Pi(x)$ is the fluctuation of the density with respect to the equilibrium value $\rho_0$, its integral vanishes, so the interaction becomes

$$V = \frac{g_0}{2\pi^2} \int dx [\partial_x \phi(x)]^2. \tag{2.188}$$

Since we have only two fields, we have only two degrees of freedom regarding renormalized parameters, and we only need to include $u$ and $K$. Once more, from dimensional analysis

$$V = \frac{\hbar}{2} \int dx \frac{u}{\pi K}[\partial_x \phi(x)]^2. \tag{2.189}$$

Putting this all together, we have the effective low-energy Hamiltonian

$$H = \frac{\hbar}{2\pi} \int dx \left[ uK(\partial_x\theta(x))^2 + \frac{u}{K}(\partial_x\phi(x))^2 \right]. \tag{2.190}$$

Note that the choice of parameters above is not restrictive, and we may as well call $\lambda_1 = uK$ and $\lambda_2 = u/K$. However, $u$ turns out to be the speed of excitations, and $K$—the Luttinger parameter—a parameter that controls long-distance correlation functions. Hence, the choice in equation (2.190) will prove to be convenient.

We now diagonalize Hamiltonian (2.190), and show that $u$ corresponds with the speed of excitations. Afterwards, we shall calculate the simplest correlation functions and relate them to the Luttinger parameter $K$. To diagonalize Hamiltonian (2.190), we write down the operators $\theta$ and $\partial_x \phi$ in terms of bosonic creation and annihilation operators $b_q^\dagger$ and $b_q$ (in the momentum representation). To do this, we use the fact that $\Pi(x) = -\partial_x \phi(x)/\pi$ is the momentum conjugate to $\theta(x)$, due to their commutator (2.177). Their expansion in bosonic operators reads

$$\theta(x) = \int \frac{dq}{\sqrt{2\pi}} \left( \frac{1}{2|q|} \right)^{1/2} \left[ b_q e^{-iqx} + b_q^\dagger e^{iqx} \right], \tag{2.191}$$

$$\Pi(x) = -i \int \frac{dq}{\sqrt{2\pi}} \left(\frac{|q|}{2}\right)^{1/2} \left[ b_q e^{-iqx} - b_q^\dagger e^{iqx} \right]. \tag{2.192}$$

The commutator is easily evaluated, and yields the correct result $[\theta(x), \Pi(x')] = i\delta(x - x')$. After some lengthy but straightforward algebra, the Hamiltonian takes the form

$$H = \frac{\hbar}{4\pi} \int dq\,|q|\left[ (v_1 - v_2)\left(b_q^\dagger b_{-q}^\dagger + b_q b_{-q}\right) + (v_1 + v_2)\left(b_q^\dagger b_q + b_q b_q^\dagger\right) \right], \tag{2.193}$$

where we have defined $v_1 = uK$ and $v_2 = \pi^2 u/K$. We clearly see that $H$ in equation (2.193) is a bosonic quadratic form and, therefore, we can diagonalize it using a Bogoliubov transformation. Before that, since the dispersion is linear and positive ($\sim|q|$), we simplify equation (2.193) to integrate over positive modes only. It becomes

$$H = \frac{\hbar}{2\pi} \int_0^\infty dq\,q\left[ (v_1 - v_2)\left(b_q^\dagger b_{-q}^\dagger + b_q b_{-q}\right) + (v_1 + v_2)\left(b_q^\dagger b_q + b_{-q}^\dagger b_{-q}\right) \right], \tag{2.194}$$

where we have dropped an irrelevant, infinite constant. To diagonalize the Hamiltonian, we use the following Bogoliubov transformation to new bosonic operators $\beta_q$ and $\beta_q^\dagger$,

$$\beta_q = \cosh \gamma\, b_q + \sinh \gamma\, b_{-q}^\dagger. \tag{2.195}$$

Setting to zero all non-diagonal terms ($\sim\beta_q^\dagger \beta_{-q}^\dagger$, $\beta_q \beta_{-q}$), implies the following relation

$$\tanh(2\gamma) = \frac{v_1 - v_2}{v_1 + v_2}, \tag{2.196}$$

and the diagonal form of the Hamiltonian becomes

$$H = \frac{\hbar}{\pi} \int_0^\infty dq\,q\sqrt{v_1 v_2}\left(\beta_q^\dagger \beta_q + \beta_{-q}^\dagger \beta_{-q}\right), \tag{2.197}$$

where, once more, we have dropped an irrelevant infinite constant. Now we can see that the spectrum, $\hbar\omega(q) = (\hbar/\pi)\sqrt{v_1 v_2}\,q$ ($q > 0$) only depends on $u$. Using $v_1 = uK$ and $v_2 = \pi^2 u/K$, we immediately see that

$$\hbar\omega(q) = \hbar u q, \quad q > 0. \tag{2.198}$$

This is exactly what we anticipated. Now we proceed to calculate some simple correlation functions and show that correlations only depend on the Luttinger parameter $K$. First, just as with the usual bosonization, we see that the parameter $\gamma$ in the Bogoliubov transformation is related to $K$ as

$$e^{2\gamma} = \frac{K}{\pi}. \tag{2.199}$$

Now, let us calculate the density-density correlation function in the ground state (vacuum), for the smeared density. This simplifies calculations and give the leading order asymptotics.

$$\langle \rho(x)\rho(0) \rangle = \langle (\rho_0 + \Pi(x))(\rho_0 + \Pi(0)) \rangle = \rho_0^2 + \rho_0 \langle \Pi(0) + \Pi(x) \rangle + \langle \Pi(x)\Pi(0) \rangle. \quad (2.200)$$

We use the Fourier decomposition of the momentum $\Pi(x)$, equation (2.192), and the inverse of the Bogoliubov transformation, given by

$$b_q = \cosh \gamma \beta_q - \sinh \gamma \beta^\dagger_{-q}, \quad (2.201)$$

obtaining, after tedious but straightforward algebra,

$$\langle \rho(x)\rho(0) \rangle = \rho_0^2 - \frac{K}{2\pi^2 x^2}. \quad (2.202)$$

Since this is a low-energy theory, the above relation (for the smeared densities) is valid as $x \to \infty$. From equation (2.202), we obtain for the pair correlation function at long distances

$$g(x) \to 1 - \frac{K}{2\pi^2 (\rho_0 x)^2}, \quad x \to \infty. \quad (2.203)$$

From it, the static structure factor takes the form

$$S(q) = \frac{K}{2\pi \rho_0}|q|, \quad q \to 0. \quad (2.204)$$

Now we have both large-distance (low-energy) and short-distance (high-energy) asymptotics for the simplest of correlation functions, if we add to Luttinger Liquid theory the theory of short-range correlations studied in the previous chapter. And, in fact, the relevant quantities, the contact and the Luttinger parameter, are related to each other. This becomes even more powerful for non-relativistic particles, which is the subject of the following section.

### 2.6.4 Non-relativistic gases as Luttinger liquids

Let us consider a non-relativistic system of spinless fermions in one spatial dimension. The low-energy physics of the problem occurs for momenta around the two Fermi points $\pm k_F$. Around these, the Galilean dispersion becomes approximately linear, since

$$\varepsilon(k) = \frac{\hbar^2 k^2}{2m} = \frac{\hbar^2 k_F^2}{2m} \pm \hbar v_F(k \mp k_F) + \frac{\hbar^2}{m}(k \mp k_F)^2, \quad (2.205)$$

where $v_F = \hbar k_F$ is the Fermi velocity. Establishing an effective theory for non-relativistic fermions consists of linearizing the dispersion, i.e., dropping the term $\propto (k \mp k_F)^2$ in equation (2.205), and considering two species of fermions, one of them right-moving and the other one left moving. Since, moreover, we would like to get rid of a *priori* irrelevant scales, we allow $k$ to take on any value in $(-\infty, \infty)$, although, as we already know, a cutoff might be eventually required. The non-interacting part of the effective Hamiltonian becomes

$$H_0 = \hbar v_F \sum_k k \left[ c_{k+}^\dagger c_{k+} - c_{k-}^\dagger c_{k-} \right] + C, \qquad (2.206)$$

with $C$ an irrelevant, infinite constant. Above, $c_{k\pm}$ are fermionic annihilation operators for right- ($+$) and left-moving ($-$) fermions.

We must now deal with interactions. Clearly, the most important interactions take place between a right-mover and left-mover with respective momenta near the Fermi points $\pm k_F$. In particular, the elastic right-left interaction takes the usual form

$$V_2 = \frac{g_2}{L} \sum_{kk'q} c_{k+q,\,+}^\dagger c_{k'-q,\,-}^\dagger c_{k'-} c_{k+}, \qquad (2.207)$$

where $g_2$ is a coupling constant. Notice how the Hamiltonian $H_0 + V_2$ is nothing but Thirring's model, with $\hbar v_F \equiv 1$ and $g_2 \equiv g$. We may also consider the so-called 'intrabranch' interactions between fermions in the same branch (right- or left-moving). These are typically included as a scattering channel. However, *there is no scattering between chiral fermions* and this 'interaction', instead, should appear as a renormalization of the non-interacting, or bare, Fermi velocity. The reason for the absence of scattering is very simple. Total momentum and total energy conservation within the same branch are equivalent to one another: there is huge degeneracy which can be dealt with by using flat-band scattering theory (see reference [15]). So let us figure out what this renormalization should look like. To do this, there are two options. The first one consists of taking into account Galilean relativity; the second one consists of a simple comparison between the leading order correction to the speed of sound for a model with known results. We will use the first method.

For a non-relativistic (Galilean) many-body system, Luttinger liquid theory is nothing but a low-energy description. Let us take the only possible low-energy theory, which is simply hydrodynamics. The classical energy of the system is the sum of kinetic and internal energy which, after defining the local density $\rho(x)$, and local density fluctuation $\delta\rho(x) \equiv \rho(x) - \rho_0$, where $\rho_0$ is the mean density, is given by

$$H[\rho(x)] = \int dx \left[ \frac{1}{2} m v(x)^2 \rho(x) + \mathcal{E}(\rho(x)) \right], \qquad (2.208)$$

where $v(x)$ is the local velocity field, and $\mathcal{E}(\rho)$ is the internal energy density, which follows the equation of state. We expand the latter around its equilibrium value—the mean density $\rho_0$—obtaining

$$\mathcal{E}(\rho(x)) = \mathcal{E}(\rho_0) + \mathcal{E}'(\rho_0)\delta\rho(x) + \frac{1}{2}\mathcal{E}''(x)[\delta\rho(x)]^2 + \mathcal{O}(\delta\rho^3). \qquad (2.209)$$

The (field-theoretical, not to be confused with the the usual isothermal compressibility, which has an extra factor of $\rho_0^2$) compressibility, $\kappa$, is related to the equation of state via

$$\kappa^{-1} = \mathcal{E}''(\rho_0), \tag{2.210}$$

and therefore at low energies

$$H[\rho(x)] = \int dx \left[ \frac{1}{2} mv(x)^2 \rho_0 + \frac{1}{2\kappa} (\delta\rho(x))^2 \right], \tag{2.211}$$

where we have dropped an irrelevant constant, and replaced $\rho(x)$ with its equilibrium value $\rho_0$. Let's now compare this with Luttinger liquid effective theory. Before proceeding, we restore physical constants ($\hbar$ and $m$). We will then use $\delta\rho(x) = -\pi^{-1}\partial_x\phi(x)$ and $v(x) = (\hbar/m)\partial_x\theta(x)$, and obtain

$$H = \frac{1}{2} \int dx \left[ \frac{m^2 uK}{\pi\hbar} v(x)^2 + \frac{\pi\hbar u}{K} (\delta\rho(x))^2 \right]. \tag{2.212}$$

Identifying the kinetic terms in both hydrodynamic and Luttinger Hamiltonians, we find that

$$\frac{u}{v_F} = \frac{1}{K}, \tag{2.213}$$

valid for every Galilean sytem. The same result is obtained by matching the interacting part.

### 2.6.5 Connecting Luttinger liquid theory and short-range universality

As a simple application, we consider the Lieb–Liniger model. We can infer the low-momentum behaviour of the static structure factor (or the long-range behaviour of the pair correlation) from its high-momentum behaviour and vice versa. To see this, take equation (1.126), which relates the high-momentum behaviour of $S(k)$ to the contact $C$ and the scattering length, as $s \equiv \lim_{k\to\infty} k^2(S(k) - 1) = 4C/Na$. Using the adiabatic theorem, equation (1.123), we see that $C = (m/\hbar^2)dE/d(-1/a)$, obtaining for the high-momentum limit, after some simple changes of variables

$$s = \frac{4m}{\hbar^2\rho} \frac{d\mathcal{E}}{d\rho} = \frac{4m}{\hbar^2\rho}\mu, \tag{2.214}$$

where $\mathcal{E} = E/L$ is the energy density, and $\mu$ is the chemical potential. For low momenta, the structure factor behaves as (see equation (2.204)) $\tilde{s} \equiv \lim_{k\to 0} S(k)/|k| = K/2\pi\rho$. Since the Lieb–Liniger model is Galilean, we have the relation $u/v_F = 1/K$, equation (2.213), between the speed of excitations and the Luttinger parameter. After some tedious algebra, we obtain the relation

$$\frac{d}{d\lambda}(\lambda s) = \frac{\hbar^2}{\tilde{s}^2}. \tag{2.215}$$

As a test, let us see that the relation works for the second-order Bethe ansatz solution to the Lieb–Liniger model. The energy per particle is given by the hard-rod expression (albeit with $a < 0$), equation (2.39), and the contact by equation (2.48). We obtain, for both the right- and left-hand sides of equation (2.215),

$$\frac{\hbar^2}{\bar{s}^2} = \frac{8mE^{(0)}}{N} \frac{\lambda(\lambda + 2)}{(1 - \lambda)^4} = \frac{d}{d\lambda}(\lambda s), \tag{2.216}$$

which, even for the hard-rod model, is exact to all orders.

## 2.7 Liberating the independent pair approximation from the Fermi sea

Here, we are going to consider a very interesting approximation that allows for non-perturbative estimation of the low-energy properties of one-dimensional systems using only two-body information, and without the Bethe ansatz. The idea is based on some very old methods pertaining to what is known in nuclear physics as the independent pair approximation (IPA) [16]. The IPA, in a nutshell, begins with a filled Fermi sea (so it only works for fermions, of course) which, to simplify things further, we assume is composed of spin-$1/2$ fermions, with balanced spin populations ($N_\uparrow = N_\downarrow = N/2$). We further assume that only $s$-wave interactions are relevant, implying that only opposite-spin pairs interact. We take each possible interacting pair within the Fermi sea, which we label by their single-particle momenta $\mathbf{k}$ and $\mathbf{k}'$, with $|\mathbf{k}|, |\mathbf{k}'| < k_F$. We let them collide in a finite volume, but we do not allow, in the collision integrals, any occupation of other modes in the Fermi sea except for their own single-particle modes $\mathbf{k}$ and $\mathbf{k}'$. We then sum up all energy shifts in this finite volume and normalize according to the number of pairs. It can be proven [16] that this procedure generates correct energy up to second-order in perturbation theory. However, it is a non-perturbative method and can have some validity for stronger interactions.

The IPA is a really powerful method. However, (i) it can be complicated to solve and, worse, (ii) it produces infrared (IR) divergences in one spatial dimension! (this is due to the IR *physical* cutoff $k_F$ and are not removable in any consistent manner). Therefore, we need another method if we want to emulate the success of the IPA in 1D. What we are going to do is remove the Fermi sea altogether, and work with two-body subsystems (all possible pairs, like in the IPA) at finite densities. Finite size effects, instead of being detrimental to the calculation, will be used to our advantage! This method is *a priori* extendable, and is one alternative way of approaching the thermodynamic limit, from a completely few-body point of view.

We begin by taking two non-relativistic spinless bosons (the fermionic solution will come from asymptotic duality) on an infinitely long line. The Hamiltonian is given by

$$H = -\frac{\hbar^2}{2m}\left[\partial_{x_1}^2 + \partial_{x_2}^2\right] + V(x_1 - x_2). \tag{2.217}$$

We assume that $\lim_{x \to \infty} x^{2-\varepsilon}V(x) = 0$ for any $\varepsilon > 0$ (see chapter 1), so that we can consider $V(x)$ a short-range interaction. We further assume that $H$ supports no two-body bound states. We solve the two-body problem by whichever means available, and extract the two-boson phase shifts $\theta(k)$ at relative momentum $k = (k_1 - k_2)/2$ which enter the scattering wave functions as

$$\psi_{K,k}(X, x) \to e^{iKX} \sin(k|x| + \theta(k)), \quad k|x| \to \infty, \tag{2.218}$$

where $K = k_1 + k_2$ is the total momentum, $X = (x_1 + x_2)/2$ is the centre of mass and $x = x_1 - x_2$ the relative coordinate. If $V(x)$ has tails, then if we place the system on a ring with length $L$, the two-body eigenenergies will necessarily contain corrections due to the long-range part of the interaction. However, in the gas phase, our target many-body system's equation of state (and therefore the energy per particle) can only depend on the $n$-body $S$-matrices, besides other scales in the problem (such as mass, Planck's constant and density), due to a theorem of Dashen, Ma and Bernstein [17]. This means that the corrections due to the tails play no role, so we need the phase shifts. This is also in accordance with asymptotic Bethe ansatz theory [3] for integrable models. Since we are going to drop all other information about the two-body problem, we can immediately assume that whatever works for bosons also works for fermions, since their asymptotic wave functions behave as

$$\phi_{K,k}(X, x) \to e^{iKX} \text{sgn}(x)\sin(k|x| + \theta(k)), \quad k|x| \to \infty, \tag{2.219}$$

and the effective range expansions and analytic properties of their phase shifts are identical for bosons and fermions (see volume 1).

Recall that, in volume 1, we investigated thoroughly how the ground state energy of a two-body system in a finite three-dimensional cube is shifted by the presence of interactions. In one spatial dimension, the result is far simpler. Take equation (2.218) and use periodic boundary conditions. We obtain the following quantization conditions,

$$kL = n\pi - 2\theta(k), \quad n \in \mathbb{Z}_+ - \{0\}, \tag{2.220}$$

$$KL = 2\pi m - \pi n, \quad m \in \mathbb{Z}. \tag{2.221}$$

For instance, in the sector of vanishing total momentum ($K = 0$), the quantum number $n = 2m$, in which case we obtain

$$k = \frac{2\pi m}{L} - \frac{2\theta(k)}{L}, \quad m \in \mathbb{Z}_+ - \{0\}, \tag{2.222}$$

which is equivalent to the single-particle quantization condition.

The quantization conditions above, equations (2.220, 2.221), imply that not every momentum scale is available from exact (numerical) diagonalization. But since we can calculate the phase shifts very efficiently using the Lippmann–Schwinger equation, that is not a problem. We now analytically continue the values of $n$ to the reals, so that equation (2.220) becomes

$$k = \frac{\pi x}{L} - \frac{2\theta(k)}{L}, \quad x \in \mathbb{R}. \tag{2.223}$$

Since we have only two particles, we identify $2/L$ with the density $\rho$, and $\pi/L$ with $k_F/2$. Equation (2.223) becomes

$$k = \frac{k_F x}{2} - \rho\theta(k) = \frac{k_F x}{2} - \frac{k_F}{\pi}\theta(k). \tag{2.224}$$

We define the variable $y = k/k_F$, and the above equation simplifies to

$$y = \frac{x}{2} - \pi^{-1}\theta(k_F y). \tag{2.225}$$

The phase shift $\theta$ is dimensionless. Therefore, it can only depend on $k_F$ via its products with other length scales in the problem, such as the scattering length and effective range. Given a particular interaction potential $V$, these length scales are completely fixed by scattering theory. If we define the (*a priori* infinite) set of length scales as

$$\mathcal{L} \equiv \{\ell_0, \ell_1, \ldots\}. \tag{2.226}$$

then the phase shift can be written as a function of these, that is

$$\theta(k_F y) = F_\theta(y; k_F\ell_0, k_F\ell_1, \ldots) = F_\theta(y; s_0, s_1, \ldots) \equiv F_\theta(y; \mathbf{s}), \tag{2.227}$$

where $s_j \equiv k_F\ell_j$ $(j = 0, 1, \ldots)$ are the dimensionless coupling constants of the theory. Once the density is fixed, the coupling constants $\mathbf{s}$ are fixed parameters, so equation (2.225) becomes

$$y = \frac{x}{2} - \pi^{-1}F_\theta(y; \mathbf{s}). \tag{2.228}$$

Before continuing with the general theory, let us solve equation (2.228) for two very simple and integrable models. First, let us consider the hard-rod model, for which $\theta(k) = -ka$. The only length scale is the hard-rod diameter $a$, so the only coupling constant of the theory is $s_0 = k_F a$. The function $F_\theta(y; s_0) = -k_F ya = -ys_0$. Therefore, equation (2.225) is immediately solved to yield

$$y = y(x) = \frac{x/2}{1 - s_0/\pi}. \tag{2.229}$$

For the case of the Calogero–Sutherland (CS) model, for which $\theta(k) = c\,\mathrm{sgn}(k)$, with $c$ a dimensionless constant, we have no length scales available, so $F_\theta(y) = c\,\mathrm{sgn}(y)$, so

$$y(x) = \frac{x/2}{1 + c\,\mathrm{sgn}(x)/\pi}. \tag{2.230}$$

Now the question is, what information about the *many-body* system can we extract with only this two-body information but without the full Bethe ansatz equations? To answer this, let us begin with $x = 2$, so that $k_F = 2\pi/L = \pi\rho$, and weakly perturb $x \to 2 + \varepsilon, \varepsilon > 0$. For the hard-rod (HR) and CS models, we obtain

$$y_{HR}(2 + \varepsilon) = \frac{1 + \varepsilon/2}{1 - s_0/\pi} = y_{HC}(2) + \frac{\varepsilon/2}{1 - s_0/\pi}, \quad (2.231)$$

$$y_{CS}(2 + \varepsilon) = y_{CS}(2) + \frac{\varepsilon/2}{1 + c/\pi}. \quad (2.232)$$

Note how we have kept $s_0$ fixed while moving the effective Fermi momentum by $2\pi\varepsilon/L$ in the case of the HR model, while in the CS model—which is scale invariant —the coupling constant naturally doesn't move. In the spirit of the IPA, but in contrast with it, we emulate the finite density by having recognised $\rho = 2/L$, and by calculating $y$ around $x = 2$, we are calculating the low-energy excitations near $k_F$. This is the case for $K = 0$, in which exciting a two-particle–two-hole pair we simply excite them from $\pm k_F$ to $\pm k_F \pm q/2$, such that the excitation energy corresponds with $\hbar\omega(q) = \varepsilon(2 + q/k_F) - \varepsilon(2)$, where

$$\varepsilon(x) = \frac{\hbar^2 k_F^2}{m} y^2(x). \quad (2.233)$$

To see how this works, let us apply it to the HR and CS models. For HRs, we have

$$\hbar\omega(q) = \frac{\hbar v_F}{(1 - \rho a)^2} q + \frac{\hbar^2 q^2}{4m(1 - \rho a)^2}. \quad (2.234)$$

From the above relation, we identify the speed of sound $v_{HR}$ and effective mass $m_{HR}$ for the excitations, obtaining

$$\frac{v_{HR}}{v_F} = \frac{1}{(1 - \rho a)^2}, \quad (2.235)$$

$$\frac{m_{HR}}{m} = (1 - \rho a)^2. \quad (2.236)$$

Because this model is extremely simple, it turns out that the speed of sound and effective mass are, in fact, exact! This can be seen by solving the Bethe ansatz integral equation (see problem 3). For the CS model, we should expect a similar feat. The excitation spectrum becomes

$$\hbar\omega(q) = \hbar v_F (1 - c/\pi) q + \frac{\hbar^2 q^2}{4m}, \quad (2.237)$$

so that the effective mass $m_{CS} = m$ remains unrenormalized while the speed of sound becomes

$$\frac{v}{v_F} = 1 - \frac{c}{\pi}. \quad (2.238)$$

Once more, for the particle-like excitations of this model, both the speed of sound and effective mass are exact. This will not remain true for any other model (integrable or otherwise), but will give a good approximation in many instances to the excitation spectrum up to and including the effective mass. To summarize the method, now in general, the speed of sound ($v$) and effective mass ($m_*$) are obtained within this approximation by calculating

$$\hbar\omega(q) = \varepsilon(2 + q/k_F) - \varepsilon(2) = \hbar v q + \frac{\hbar^2 q^2}{4m_*} + \cdots. \tag{2.239}$$

We also note that the phenomenological expression from non-linear Luttinger liquid theory (reference [18]), which has been proven for Bethe ansatz-solvable models, is given by

$$\frac{m}{m_*} = \frac{1}{2vK_L^{1/2}} \frac{d(\rho v)}{d\rho}. \tag{2.240}$$

An interesting property of this theory is the following

**Theorem 1** *For a non-relativistic, Galilean system within the independent pair approximation without a Fermi sea, the speed of sound and the Luttinger parameter are given by the Fermi velocity $v_F$ and $K_L = 1$, respectively, if the two-body phase shift $\theta(k_F) = 0$, and $\theta(k)$ is continuous and differentiable at $k = k_F$.*

The proof of the above theorem is straightfoward. Note that $K_L = 1$ does *not* mean the system is the Tonks–Girardeau gas nor a free Fermi gas. It means that interactions from opposite Fermi points have vanishing phase shifts, just as free fermions or Tonks–Girardeau bosons. In general, the theorem does not apply for non-integrable models. And in known integrable models this only happens if the model is free fermions or Tonks–Girardeau bosons. However, it is a very good approximation in models with realistic interactions (see reference [19]), and coincides, of course, with constructive Luttinger liquid predicitions, although in that case it reduces once more to free fermions. Deviations from this rule, for non-integrable models, can be attributed to three- and higher-body interaction effects.

## 2.8 Trapped multicomponent systems

In this section, we will finally consider trapped multicomponent systems. We will focus on spin-1/2 fermions, which are the simplest case, and is sufficiently illustrative for our purposes. We follow reference [20].

Consider a generic system of $N$ equal mass particles in an external trap, interacting via LO even-wave interactions. Their Hamiltonian is given by

$$H = \sum_{i=1}^{N} \left( \frac{p_i^2}{2m} + U(x_i) \right) + g \sum_{i<j=1}^{N} \delta(x_i - x_j). \tag{2.241}$$

The non-interacting Hamiltonian is not homogeneous, as it contains an external trapping potential $U(x)$. The single-particle eigenvectors of the non-interacting Hamiltonian are denoted $\psi_n(x)$, and their associated eigenenergies $E_n$. We shall construct many-body, spin-1/2 fermionic eigenstates of Hamiltonian (2.241) in the limit $1/g \to 0^+$ from Slater determinants with non-interacting orbitals $\psi_n(x_i)$. The stationary Schrödinger equation with hard core interactions has solutions which, for each permutation sector (say, $P1 = \{x_1 < x_2 < ... < x_N\}$) correspond to Slater determinants. That is, regardless of symmetry and statistics, the eigenstates take the form for $1/g = 0$

$$\psi_P(x_1, ...,x_N) \propto \mathcal{A}(\psi_{n_1}(x_1) ... \psi_{n_N}(x_N)), \qquad (2.242)$$

in whichever permutation sector $P$. Above, $\mathcal{A}$ is the antisymmetrization operator. As an extremely simple example, take two bosons. In the sector $x_1 < x_2$, we have $\psi(x_1, x_2) = \psi_{n_1}(x_1)\psi_{n_2}(x_2) - \psi_{n_2}(x_1)\psi_{n_1}(x_2)$. In the sector $x_1 > x_2$, we have $\psi(x_1, x_2) = \lambda[\psi_{n_1}(x_1)\psi_{n_2}(x_2) - \psi_{n_2}(x_1)\psi_{n_1}(x_2)]$. Now, take $x < y$, and because of bosonic symmetry, $\psi(x, y) = \psi(y, x)$, which implies that $\lambda = -1$. That is, in *any* sector, the bosonic eigenstates are given by

$$\psi(x_1, x_2) = \mathcal{A}[\psi_{n_1}\psi_{n_2}]\Theta[x_2 - x_1] - \mathcal{A}[\psi_{n_1}\psi_{n_2}]\Theta(x_1 - x_2) \propto \text{sgn}(x_1 - x_2)\mathcal{A}[\psi_{n_1}\psi_{n_2}]. \quad (2.243)$$

The last term is the usual form in which we have been writing Girardeau's wave functions. Now that this is clear, we can write the general form for an $N$-body wave function regardless of symmetry

$$\psi(x_1, x_2, ...,x_N) = \sum_P \lambda_P \mathcal{A}[\psi_{n_1}\psi_{n_2} ... \psi_{n_N}]\chi(\mathbf{x} \in P), \qquad (2.244)$$

where the sum is over all possible permutations, $\lambda_P$ are constant coefficients (with signs that depend on statistics and amplitudes to be calculated), and $\chi(\mathbf{x} \in P)$ is a characteristic function that vanishes unless $\mathbf{x} \in P$, in which case $\chi = 1$.

Let us consider spin-1/2 fermions. The wave function, for finite $g$, to the Lieb–Liniger boundary condition

$$\frac{1}{2g}[(\partial_{x_i}\psi - \partial_{x_j}\psi)_{x_{ij} \to 0^+} - (\partial_{x_i}\psi - \partial_{x_j}\psi)_{x_{ij} \to 0^-}] = \psi(x_i = x_j). \qquad (2.245)$$

Since both the left- and right-hand-sides vanish as $g \to 0$, we need a device to do perturbation theory at the limit $g \to 0$. The simplest way is to use the Hellmann–Feynman theorem. If $\psi$ is an eigenstate with energy $E(g)$, with $g$ finite, then

$$Q \equiv -\lim_{g \to \infty} \frac{\partial E}{\partial g^{-1}} = \lim_{g \to \infty} g^2 \frac{\sum\limits_{i<j=1}^{N} \int \prod\limits_{k=1}^{N} dx_k |\psi|^2 \delta(x_i - x_j)}{||\psi||^2}. \qquad (2.246)$$

Using the boundary conditions (2.245) in the equation above, we obtain

$$Q = \frac{\sum\limits_{i<j=1}^{N} \int \prod\limits_{k=1}^{N} dx_k \delta(x_i - x_j)\big|(\partial_{x_i}\psi - \partial_{x_j}\psi)_{x_{ij} \to 0^+} - (\partial_{x_i}\psi - \partial_{x_j}\psi)_{x_{ij} \to 0^-}\big|^2}{||\psi||^2} \qquad (2.247)$$

The integral above is divided into sectors as

$$Q = \frac{\sum_{i<j=1}^{N}\sum_{k=1}^{N!}\int_{\Gamma_k}\int\prod_{i=1}^{N} dx_i\,\delta(x_i - x_j)\left|(\partial_{x_i}\psi - \partial_{x_j}\psi)_{x_{ij}\to 0^+} - (\partial_{x_i}\psi - \partial_{x_j}\psi)_{x_{ij}\to 0^-}\right|^2}{||\psi||^2}, \tag{2.248}$$

where $\Gamma_k$ are permutation sectors. Using the above relation we obtain, after straightforward algebra,

$$Q = \frac{\sum_{k,p}(\lambda_k - \lambda_p)^2\alpha_{kp}}{\sum_k\lambda_k^2}. \tag{2.249}$$

Above, $k$ and $p$ run over the number of independent coefficients. In the case of spin-1/2, this number is $M = N!/(N_\uparrow!N_\downarrow!)$. The relevant couplings, $\alpha_{kp}$, in equation (2.249), are given by

$$\alpha_{kp} = \int_{\Gamma_k} dx_1 \dots dx_N\,\delta(x_i - x_j)|\partial_{x_i}\psi_A|^2, \tag{2.250}$$

where $\psi_A = \mathcal{A}[\psi_{n_1}\psi_{n_2}\dots\psi_{n_N}]$. A very powerful algorithm (called CONAN) for obtaining these coefficients and solving the problem is given in reference [21].

## Problems

1. Consider the interaction in equation (2.79) for two fermions. Show that if the interaction strength is made $\propto b^{-(2+\varepsilon)}$, for $\varepsilon > 0$, an appropriate choice of $\varepsilon$ allows for fine tuning of the odd-wave scattering length to any target value as $b$ approaches 0, without arriving at the exact limit.
2. Show that the Gaudin's magnet Hamiltonian, equation (2.97), can be recast, except for a constant offset, as equation (2.98).
3. Obtain the speed of sound for the HR model by solving the ground state Bethe ansatz equation. (Hint: calculate the ground-state chemical potential and use $mv^2 = \rho\partial_\rho\mu$.)
4. Calculate the ground state energy of the attractive Lieb–Liniger model. (Hint: it is a simple counting problem.)

## References

[1] Girardeau M 1960 Relationship between systems of impenetrable bosons and fermions in one dimension *J. Math. Phys.* **1** 516
[2] Lieb E H and Liniger W 1963 Exact analysis of an interacting Bose gas. I. The general solution and the ground state *Phys. Rev.* **130** 1605
[3] Sutherland B 2004 *Beautiful Models: 70 Years of Exactly Solved Quantum Many-body Problems* (Singapore: World Scientific)
[4] Valiente M 2025 Exact thermal distributions in integrable classical and quantum gases arXiv:2503.07141

[5] Slavnov N 2022 *Algebraic Bethe Ansatz and Correlation Functions* (Singapore: World Scientific)

[6] McGuire J B 1964 Study of exactly soluble one-dimensional N-body problems *J. Math. Phys.* **5** 622

[7] Shirokov Y M 1976 Combined algebra for quantum and classical mechanics *Teor. Mat. Fiz. (USSR)* **28** 806–13

[8] Sekino Y and Nishida Y 2021 Field-theoretical aspects of one-dimensional Bose and Fermi gases with contact interactions *Phys. Rev.* A **103** 043307

[9] Holten M *et al* 2022 Observation of Cooper pairs in a mesoscopic two-dimensional Fermi gas *Nature* **606** 287

[10] Dukelsky J, Pittel S and Sierra G 2004 Colloquium: exactly solvable Richardson-Gaudin models for many-body quantum systems *Rev. Mod. Phys.* **76** 643

[11] Claeys P W 2018 *PhD Thesis: Richardson–Gaudin models and broken integrability* arXiv:1809.04447

[12] Landau L D 1957 The theory of a fermi liquid *Sov. Phys. JETP* **3** 920

[13] Haldane F D M 1981 Effective harmonic-fluid approach to low-energy properties of one-dimensional quantum fluids *Phys. Rev. Lett.* **47** 1840

[14] Giamarchi T 2003 *Quantum Physics in One Dimension* (Oxford: Oxford University Press)

[15] Valiente M and Zinner N T 2017 Quantum collision theory in flat bands *J. Phys. B: At. Mol. Opt. Phys.* **50** 064004

[16] Fetter A and Walecka J D 2003 *Quantum Theory of Many-Particle Systems* (New York: Dover)

[17] Dashen R, Ma S-K and Bernstein H J 1969 S-matrix formulation of statistical mechanics *Phys. Rev.* **187** 345

[18] Imambekov A, Schmidt T L and Glazman L I 2012 One-dimensional quantum liquids: beyond the Luttinger liquid paradigm *Rev. Mod. Phys.* **84** 1253

[19] Valiente M and Öhberg P 2016 Few-body route to one-dimensional quantum liquids *Phys. Rev.* A **94** 051606

[20] Volosniev A *et al* 2014 Strongly interacting confined quantum systems in one dimension *Nat. Commun.* **5** 5300

[21] Loft N J S, Kristensen L B, Thomsen A E, Volosniev A G and Zinner N T 2016 CONAN– the cruncher of local exchange coefficients for strongly interacting confined systems in one dimension *Comput. Phys. Commun.* **209** 171

**IOP** Publishing

Strongly Interacting Quantum Systems, Volume 2
Many-body physics
**Manuel Valiente and Nikolaj Thomas Zinner**

# Chapter 3

# Virial expansions

Low-temperature thermodynamics in strongly interacting systems is quite complicated. This is true even for classical statistical mechanics, except in simple systems and extremely close to zero temperature, where the system simply stays in the minimal potential energy spatial configuration. For high temperatures, there exists a method (as old as or older than statistical mechanics itself) called the virial expansion. This allows for the calculation of a series expansion for small fugacities. In the classical case, the expansion is somewhat simple and, in fact, it is nowadays of little use due to the power of current computers: we can solve huge, strongly interacting, classical systems exactly at any temperature and under arbitrary external conditions. In the quantum case, this is not so easy. While quantum Monte Carlo has gone a long way, and density-matrix-renormalization group (DMRG) calculations (in one spatial dimension) are essentially exact and efficient, the former still suffers from the same old sign problem, and DMRG is restricted to low-entanglement (low temperatures) and low dimensionality. Therefore, quantum virial expansions can be very powerful for, at least, predicting some mid-to-high-temperature properties regardless of whether the system is weakly or strongly interacting. Although we shall not use them, deriving the leading order cluster expansions for classical systems is useful and inspiring for the quantum case, so we shall begin with them.

## 3.1 Classical virial expansion

We consider a three-dimensional system of $N$ identical, classical, non-relativistic point particles interacting via pairwise potentials, with Hamiltonian

$$H(p, q) = \sum_{i=1}^{N} \frac{\mathbf{p}_i^2}{2m} + \sum_{i<j=1}^{N} V(\mathbf{r}_i - \mathbf{r}_j).$$

(3.1)

doi:10.1088/978-0-7503-3091-6ch3

We are interested in calculating a low-fugacity expansion for the canonical partition function $Z(\beta, N, \mathcal{V})$, with $\beta = 1/k_B T$ the inverse temperature and $\mathcal{V}$ the volume. The partition function is defined, as usual, via

$$Z(\beta, N, V) = \frac{1}{N!} \int_{\mathbb{R}^{3N}} \prod_{i=1}^{N} \left(\frac{d\mathbf{p}_i}{2\pi\hbar}\right) \int_{\mathcal{V}^N} \prod_{j=1}^{N} d\mathbf{r}_j e^{-\beta H(p, q)}. \qquad (3.2)$$

For a non-relativistic, Galilean system, we get rid of the momentum dependence by integrating over all momenta (it is a Gaussian integral). The partition function becomes

$$Z(\beta, N, \mathcal{V}) = \frac{1}{N!} \int_{\mathcal{V}^N} \prod_{i=1}^{N} \left(\frac{d\mathbf{r}_i}{\lambda_T^3}\right) e^{-\beta \sum_{j<k=1}^{N} V(\mathbf{r}_i - \mathbf{r}_j)}. \qquad (3.3)$$

Above, $\lambda_T = (2\pi\hbar^2/mk_B T)^{1/2}$ is the thermal wavelength. Note how $Z$ in equation (3.3) is impossible to calculate exactly by means of analytic methods. This is where the virial expansion comes in.

To see how it works, let us assume that the temperature is high when compared with other typical energy scales. Mathematically, this is the limit $\beta \to 0^+$. Then, $\exp(-\beta V)$ is typically near 1, and therefore $\lambda = \exp(-\beta V) - 1$ is a small parameter ($|\lambda| \ll 1$). To be specific, let us define

$$\lambda(\mathbf{r}_i - \mathbf{r}_j) = \lambda(\mathbf{r}_{ij}) = e^{-\beta V(\mathbf{r}_i - \mathbf{r}_j)} - 1. \qquad (3.4)$$

The canonical partition function becomes

$$Z(\beta, N, \mathcal{V}) = \frac{1}{N!\lambda_T^{3N}} \int_{\mathcal{V}^N} d\mathbf{r}_1 \cdots d\mathbf{r}_N \prod_{j<k=1}^{N} [1 + \lambda(\mathbf{r}_{ij})]. \qquad (3.5)$$

The product above can be expanded as sums containing two-, three- and $n$-body terms ($3 < n \leqslant N$). Since we are only interested in illustrating the method, and the quantum case becomes very complicated for three- and higher-body terms, we cut the expansion as

$$Z(\beta, N, \mathcal{V}) \approx Z^{(2)}(\beta, N, \mathcal{V}) \equiv \frac{1}{N!\lambda_T^{3N}} \int_{\mathcal{V}^N} d\mathbf{r}_1 \cdots d\mathbf{r}_N \left[1 + \sum_{i<j=1}^{N} \lambda(\mathbf{r}_{ij})\right]. \qquad (3.6)$$

Since each term above depends only on two positions, equation (3.6) is simplified to

$$Z^{(2)}(\beta, N, \mathcal{V}) = \frac{1}{N!\lambda_T^{3N}} \left[\mathcal{V}^N + \frac{N(N-1)}{2} \mathcal{V}^{N-2} \mathcal{I}(\beta, \mathcal{V})\right], \qquad (3.7)$$

where we have defined the integral

$$\mathcal{I}(\beta, \mathcal{V}) = \int_{\mathcal{V}} d\mathbf{r}_1 \int_{\mathcal{V}} d\mathbf{r}_2 \lambda(\mathbf{r}_{12}). \qquad (3.8)$$

Since we are interested in the thermodynamic limit, we can change variables in the integral to centre of mass and relative coordinates, obtaining

$$\mathcal{I}(\beta, \mathcal{V}) \to \int_{\mathcal{V}} d\mathbf{R} \int_{\mathbb{R}^3} d\mathbf{r}\lambda(\mathbf{r}) = \mathcal{V} \int_{\mathbb{R}^3} d\mathbf{r}\lambda(\mathbf{r}). \tag{3.9}$$

For typical interactions, the above integral is convergent. In summary, the second-order (including up to two-particle interactions) partition function is given by

$$Z^{(2)}(\beta, N, \mathcal{V}) = \frac{\mathcal{V}^N}{N!\lambda_T^{3N}} \left[ 1 + \frac{N(N-1)}{2\mathcal{V}} \int_{\mathbb{R}^3} d\mathbf{r}\lambda(\mathbf{r}) \right]. \tag{3.10}$$

We further define the standard virial coefficients

$$b_1 = \frac{1}{\mathcal{V}} \int_{\mathcal{V}} d\mathbf{r}1 = 1, \tag{3.11}$$

$$b_2 = \frac{1}{2!\lambda_T^3 \mathcal{V}}\mathcal{I}(\beta, \mathcal{V}) = \frac{1}{2\lambda_T^3} \int_{\mathbb{R}^3} d\mathbf{r}\lambda(\mathbf{r}). \tag{3.12}$$

Now, to obtain an approximation to the pressure, we need to be careful regarding which quantity we expand. The second-order pressure is given by

$$P^{(2)}(T, v) = k_B T \left( \frac{\partial \ln Z^{(2)}}{\partial V} \right)_{N,T}, \tag{3.13}$$

where $v = V/N = 1/\rho$ is the reduced volume (or inverse density). Expanding for small $1/v$ (low-density), we obtain the desired expression

$$\frac{P^{(2)}v}{k_B T} = 1 - \frac{\lambda_T^3 b_2}{v}. \tag{3.14}$$

Let us now obtain the second virial coefficient for some simple models. The three-dimensional hard-sphere model with diameter $a$ is especially easy to calculate, since $\lambda(\mathbf{r}) = -\Theta(a - |\mathbf{r}|)$. The second virial coefficient is given by

$$b_2 = -\frac{1}{2\lambda_T^3} \int_{\mathbb{R}^3} d\mathbf{r}\Theta(a - |\mathbf{r}|) = -\frac{4\pi}{2\lambda_T^3} \int_0^a dr r^2 = -\frac{2\pi a^3}{3\lambda_T^3}. \tag{3.15}$$

If we do this instead in one dimension (Tonks gas), we only need to change $v \to N/L$ and $\lambda_T^3 \to \lambda_T$, obtaining

$$b_2 = -\frac{1}{2\lambda_T} \int_{-a}^a dx = -\frac{a}{\lambda_T}. \tag{3.16}$$

In this case, we can compare with the exact results, since the Tonks gas is exactly solvable. To extract the pressure, we solve the thermodynamic Bethe ansatz (see chapter 2). The canonical partition function has the form (see problem 1)

$$Z = \frac{L^N}{\lambda_T^N}[(1 - \rho a)]^N. \tag{3.17}$$

This gives for the pressure

$$\frac{Pv}{k_{\mathrm{B}}T} = 1 - \frac{a/v}{1 - a/v} \implies \frac{P^{(2)}v}{k_{\mathrm{B}}T} = 1 - \frac{a}{v}, \tag{3.18}$$

which gives $b_2 = -a/\lambda_T$, as should be.

## 3.2 Quantum virial expansion

We now move on to the more interesting quantum statistical mechanical problem. In this case, we will make use of scattering theory and energy shifts in finite volumes, explained at length in volume 1. The canonical partition function for quantum systems is given by

$$Z(\beta, N, \mathcal{V}) = \mathrm{Tr}(e^{-\beta H}), \tag{3.19}$$

where $H$ is now a Hamiltonian operator. For non-relativistic systems with pairwise interactions we simply take the quantised form of equation (3.1).

Our goal is now to evaluate, approximately, the trace of the thermal density matrix $\exp(-\beta H)$. To do this, we place an $N$-body quantum system in a finite volume $\mathcal{V}$, and define a complete orthonormal set of wave functions in the Hilbert space, $\{|\psi_\alpha\rangle\}_\alpha$, where $\alpha$ are indices that can be mapped to the natural numbers, since the Hilbert space is separable (by the principles of quantum mechanics!). The trace can then be evaluated as

$$Z(\beta, N, \mathcal{V}) = \int_{\mathcal{V}^N} \left( \prod_{i=1}^N d\mathbf{r}_i \right) \sum_\alpha \psi_\alpha^*(\mathbf{r}_1, \ldots, \mathbf{r}_N) e^{-\beta H} \psi_\alpha(\mathbf{r}_1, \ldots, \mathbf{r}_N). \tag{3.20}$$

Following Huang [1], for different particle numbers $N = 1, 2, \ldots$, we define the following function

$$W_N(\mathbf{r}_1, \ldots, \mathbf{r}_N) = N! \lambda_T^{3N} \sum_\alpha \psi_\alpha^*(\mathbf{r}_1, \ldots, \mathbf{r}_N) e^{-\beta H} \psi_\alpha(\mathbf{r}_1, \ldots, \mathbf{r}_N), \tag{3.21}$$

so that $N! \lambda_T^{3N} Z = \int_{\mathcal{V}^N} W_N$. This fixes the normalization so that $W_1(\mathbf{r}) = 1$ as $\mathcal{V} \to \infty$. To see this, recall that if $\hat{O}$ is an operator and $|\psi\rangle$ is an eigenstate, $\hat{O}|\psi\rangle = o|\psi\rangle$, then $f(\hat{O})|\psi\rangle = f(o)|\psi\rangle$. Therefore, since the one-body Hamiltonian is just $H = \mathbf{p}_1^2/2m$, if we choose the one-body orthonormal set with periodic boundary conditions in a cube with volume $\mathcal{V}$ as

$$\psi_{\mathbf{n}}(\mathbf{r}) = \frac{1}{\sqrt{\mathcal{V}}} e^{i\mathbf{k_n} \cdot \mathbf{r}}, \tag{3.22}$$

with $k_{\mathbf{n}} = 2\pi\mathbf{n}/\mathcal{V}^{1/3}$ we obtain

$$W_1(\mathbf{r}) = \frac{1}{\mathcal{V}} \lambda_T^3 \sum_{\mathbf{n} \in \mathbb{Z}^3} \exp(-\beta\hbar^2 |\mathbf{k_n}|^2/2m) \to \lambda_T^3 \int_{\mathbb{R}^3} \frac{d\mathbf{k}}{(2\pi)^3} e^{-\beta\hbar^2 |\mathbf{k}|^2/2m} = 1, \tag{3.23}$$

where the continuum limit has been taken (using, for instance, the Poisson summation formula). We now consider the two-body function $W_2(\mathbf{r}_1, ...,\mathbf{r}_N)$. The most convenient basis to calculate $W_2$ is the eigenfunction basis of the Hamiltonian in a finite volume. Just as in the classical case, we can eliminate the centre-of-mass, since the two-body eigenstates have the form

$$\psi_{\mathbf{K},\mathbf{k}}(\mathbf{R}, \mathbf{r}) = \frac{e^{i\mathbf{K}\cdot\mathbf{R}}}{\sqrt{\mathcal{V}}}\psi_{\mathbf{k}}(\mathbf{r}), \tag{3.24}$$

for scattering states (i.e., for states that becomes scattering wave functions in the limit $\mathcal{V} \to \infty$), and

$$\psi_{\mathbf{K},\mathbf{n}}(\mathbf{R}, \mathbf{r}) = \frac{e^{i\mathbf{K}\cdot\mathbf{R}}}{\sqrt{\mathcal{V}}}\psi_{\mathbf{n}}(\mathbf{r}), \tag{3.25}$$

for two-body bound states (i.e., states that remain normalizable in the relative coordinate as $\mathcal{V} \to \infty$). We immediately see that $W_2$ can be written as

$$W_2(\mathbf{R}, \mathbf{r}) = 2\lambda_T^6\left(\frac{1}{\mathcal{V}}\sum_{\mathbf{K}}e^{-\beta\hbar^2\mathbf{K}^2/4m}\right)\left[\sum_{\mathbf{n}}e^{-\beta E_{\mathbf{n}}}|\psi_{\mathbf{n}}(\mathbf{r})|^2 + \sum_{\mathbf{k}}e^{-\beta\varepsilon(\mathbf{k})}|\psi_{\mathbf{k}}(\mathbf{r})|^2\right]. \tag{3.26}$$

Above, $\varepsilon(\mathbf{k})$ is the shifted energy of the scattering state (labelled by its non-interacting limit). The sum over total momenta is easily evaluated by taking the continuum limit,

$$\left(\frac{1}{\mathcal{V}}\sum_{\mathbf{K}}e^{-\beta\hbar^2\mathbf{K}^2/4m}\right) \to \int_0^\infty \frac{dK}{(2\pi)^3}4\pi K^2 e^{-\beta\hbar^2K^2/4m} = \frac{2\sqrt{2}}{\lambda_T^3}. \tag{3.27}$$

The sum over the bound states, in general, cannot be performed analytically, but since $W_2(\mathbf{R}, \mathbf{r})$ does not depend on $\mathbf{R}$, and $\psi_{\mathbf{n}}$ is normalized, the dependence on the bound state wave function disappears. The same applies for the scattering wave functions. We now handle the shifted energies of the scattering states. Since we are interested in the large-volume limit, the actual boundary conditions for the scattering states are unimportant. If the two-body interaction is spherically symmetric, the simplest way to go is to calculate energy shifts with open boundary conditions on the surface of a sphere with radius $R_0$, such that $4\pi R_0^3/3 = \mathcal{V}$. If we do this, the quantization of $\mathbf{k}$ will not be 'Cartesian' but spherical. Using spherical coordinates for the scattering states, we have

$$\psi_{q,\ell,m}(\mathbf{r}) = u_{q\ell}(r)Y_{\ell m}(\Omega), \tag{3.28}$$

and the boundary condition reduces to $u_{q\ell}(R_0) = 0$. This condition on partial-wave scattering states implies

$$\sin(qR_0 + \pi\ell/2 + \theta_\ell(q)) = 0 \implies qR_0 = \pi p - \frac{\pi}{2}\ell - \theta_\ell(q), \tag{3.29}$$

where $p \in \mathbb{Z}_+ \cup \{0\}$ and $\theta_\ell(k)$ is the partial wave phase shift. Since the non-interacting $k$-label is given by equation (3.29) with $\theta_\ell \equiv 0$, and for $\mathcal{V} \to \infty$ we have $q \to k$, we can use for the $q$-quantization

$$q \approx k - \frac{\theta_\ell(k)}{R_0}. \tag{3.30}$$

The sum $\sum_{\mathbf{k}} f_\ell(k)$ is to be replaced as

$$\sum_{\mathbf{k}} f_\ell(k) \to \sum_{\ell=0}^{\infty} (2\ell + 1) \int_0^\infty \frac{dk}{\pi} R_0 f_\ell(k). \tag{3.31}$$

We change variables as in equation (3.30), so that equation (3.31) becomes

$$\sum_{\mathbf{k}} f_\ell(\mathbf{k}) \to \sum_{\ell=0}^{\infty} (2\ell + 1) \int_0^\infty dq \left( 1 + \frac{1}{R_0} \frac{\partial \theta_\ell}{\partial q} \right) \tilde{f}_\ell(q), \tag{3.32}$$

where $\tilde{f}_\ell(q) = f_\ell(q - \theta_\ell(q)/R_0)$. The reason why this is a helpful change of variables is because the function we are interested in is $\tilde{f}_\ell(q) = \exp(-\beta \hbar^2 q^2/m)$, so that the integral is more conveniently expressed by integrating over $q$ instead of $k$.

We now calculate the function $Z(\beta, N = 2, \mathcal{V})$, which is given by

$$
\begin{aligned}
Z(\beta, 2, \mathcal{V}) &= \frac{1}{2\lambda_T^6} \int_{\mathcal{V} \times \mathcal{V}} d\mathbf{r}_1 d\mathbf{r}_2 \, W_2(\mathbf{r}_1, \mathbf{r}_2) \\
&= \frac{1}{2\lambda_T^6} 4\sqrt{2} \lambda_T^3 \mathcal{V} \int d\mathbf{r} \left[ \sum_{\mathbf{n}} e^{-\beta E_{\mathbf{n}}} |\psi_{\mathbf{n}}(\mathbf{r})|^2 + \sum_{\mathbf{k}} e^{-\beta \varepsilon(\mathbf{k})} |\psi_{\mathbf{k}}(\mathbf{r})|^2 \right] \\
&= \frac{2\sqrt{2}}{\lambda_T^3} \mathcal{V} \left[ \sum_{\mathbf{n}} e^{-\beta E_{\mathbf{n}}} + \sum_{\mathbf{k}} e^{-\beta \varepsilon(\mathbf{k})} \right] \\
&= \frac{2\sqrt{2}}{\lambda_T^3} \mathcal{V} \left[ \sum_{\mathbf{n}} e^{-\beta E_{\mathbf{n}}} + \sum_{\ell=0}^{\infty} (2\ell + 1) \int_0^\infty \frac{dq}{\pi} \left( R_0 + \frac{\partial \theta_\ell(q)}{\partial q} \right) e^{-\beta \hbar^2 q^2/m} \right].
\end{aligned} \tag{3.33}
$$

We define the non-interacting two-body partition function as $Z_0(\beta, 2, V)$, so that equation (3.33) can be rewritten as

$$Z - Z_0 = \frac{2\sqrt{2}}{\lambda_T^3} \mathcal{V} \left[ \sum_{\mathbf{n}} e^{-\beta E_{\mathbf{n}}} + \sum_{\ell=0}^{\infty} (2\ell + 1) \int_0^\infty \frac{dq}{\pi} \frac{\partial \theta_\ell(q)}{\partial q} e^{-\beta \hbar^2 q^2/m} \right]. \tag{3.34}$$

Note that equation (3.34) is identical to the one derived by Huang [1]. However, the method followed there uses the density of states, and we find the method of changing the variables (which is exactly equivalent) to be conceptually simpler.

Let us see how to calculate the pressure, by using the grand-canonical ensemble. Denoting $z = e^{\beta \mu}$, where $\mu$ is the chemical potential, the grand-canonical partition function is given by

$$\mathcal{Z}(\beta, \mu, \mathcal{V}) = \mathrm{Tr}(e^{-\beta(H - \mu \hat{N})}) = \sum_{N=0}^{\infty} z^N Z(\beta, N, \mathcal{V}). \tag{3.35}$$

Hence, to second order in the virial expansion, we have

$$\mathcal{Z}^{(2)}(\beta, \mu, \mathcal{V}) = 1 + zZ(\beta, 1, \mathcal{V}) + z^2 Z(\beta, 2, \mathcal{V}). \qquad (3.36)$$

From our previous analysis, $W_1(\mathbf{r}) = 1$, so $Z(\beta, 1, \mathcal{V}) = \mathcal{V}\lambda_T^{-3}$. For two particles, we have equation (3.34). Rearranging the terms we obtained as

$$\frac{P^{(2)}\mathcal{V}}{k_B T} = Z(\beta, 1, \mathcal{V})\sum_{k>0} z^k b_k = \frac{\mathcal{V}}{\lambda_T^3}\sum_{k>0} z^k b_k, \qquad (3.37)$$

where $b_k$ are the so-called virial coefficients, we finally get

$$b_1 = 1, \qquad (3.38)$$

$$b_2 = Z(\beta, 2, \mathcal{V}) - \frac{1}{2}Z(\beta, 1, \mathcal{V})^2 = Z(\beta, 2, \mathcal{V}) - \left(\frac{1}{\lambda_T^3}\right)^2. \qquad (3.39)$$

## Problems

1. Consider the one-dimensional hard-rod Bose or spinless Fermi gas (they are thermodynamically equivalent). Since it is exactly solvable, we can compare the exact solution to the virial expansion. The differences between three and one dimension are the following: (i) instead of $\lambda_T^3$ we have $\lambda_T$; (ii) instead of angular momentum $\ell$ we have parity (even for bosons, odd for fermions); (iii) the boundary condition is replaced with $\sin(kL/2 + \theta_\pm(k)) = 0$, so we replace $R_0$ with $L/2$; and (iv) the integral over total momentum changes as

$$\frac{1}{L}\sum_K e^{-\beta\hbar^2 K^2/4m} \rightarrow \int_{-\infty}^{\infty} \frac{dK}{2\pi} e^{-\beta\hbar^2 K^2/4m} = \sqrt{\frac{m}{\pi\beta\hbar^2}} = \frac{\sqrt{2}}{\lambda_T}.$$

   Taking all of this in account, we have $b_1 = 1$, and we obtain $b_2$ from $Z(\beta, 2, L)$, as

$$Z = \frac{\sqrt{2}}{\lambda_T} L \left[ \sum_n e^{-\beta E_n} + \int_0^{\infty} \frac{dq}{\pi}\left(\frac{L}{2} + \frac{\partial\theta_\pm(q)}{\partial q}\right) \right].$$

   (a) Obtain the second virial coefficient for the hard-rod quantum gas of bosons or spinless fermions.
   (b) Solve the thermodynamic Bethe ansatz exactly and show that the resulting virial expansion agrees to second order.
2. Consider the Lieb–Liniger model with scattering length $a < 0$ and density $\rho$.
   (a) Obtain the second-order virial expansion for the pressure.
   (b) Compare the resulting expression with the result for hard rods of diameter $a > 0$ and density $\rho$.

3. Consider a system of three-dimensional spinless fermions interacting via a $p$-wave interaction with phase shifts $\theta_{\ell=1}(k)$. Assume there are no bound states.

    (a) Obtain a general expression for the second-order virial expansion for the pressure.

    (b) Identify the dilute limit, using the effective range expansion for the $p$-wave phase shifts, and obtain an approximate virial expansion.

# Reference

[1] Huang K 1963 *Statistical Mechanics* (New York: Wiley)

**IOP** Publishing

Strongly Interacting Quantum Systems, Volume 2
Many-body physics
**Manuel Valiente and Nikolaj Thomas Zinner**

# Chapter 4

## Few-body systems embedded in a many-body medium

So far, we have used a number of powerful few-body methods to obtain non-perturbative physics, either approximately, effectively or exactly, in many-body quantum systems. In this chapter, we treat what is commonly known as 'polaron problems'. These consist of a many-body background, which can be non-interacting, weakly interacting or strongly interacting, and one or two particles of a different species than the majority one which is thermodynamically large. Of course, the minority particles interact with the many-body background. The effect of the medium on the minority particles is to 'dress' them, due to some polarization. These are typically called 'polarons'. Another name they may come by is impurities, which can be either itinerant (mobile) or static (such as defects and vacancies). Most of the time, we shall use the convention of calling anything that moves 'polaron', and every static particle 'impurity'. We begin with the infinitely massive case of minority particles, that is, impurities, and then move on to study the cases of a polaron embedded in a Fermi sea and, finally, the effective interactions between heavy polarons in a Fermi sea of light fermions.

### 4.1 Impurities

A static, non-magnetic impurity interacting with an otherwise non-interacting Fermi gas can be modelled as a static one-particle interaction potential. That is, the many-body Hamiltonian is simply

$$H = \sum_{i=1}^{N} \frac{\mathbf{p}_i^2}{2m} + \sum_{i=1}^{N} U(\mathbf{r}_i), \tag{4.1}$$

where $U(\mathbf{r})$ is the impurity–fermion interaction. Typically, this will be short-ranged, and therefore will admit a number ($M$) of bound states and, of course, a continuum of scattering states, in the infinite volume limit. For simplicity, and without loss of

doi:10.1088/978-0-7503-3091-6ch4

generality, we will consider the mobile particles to be spinless fermions. If we place the system in a sphere of radius $R_0$ with open boundary conditions on the surface, the many-body problem reduces to a one-body problem. In the thermodynamic limit, the effect of the impurity on the energy of the system is of $O(1)$ (not extensive), since single-particle energy shifts are of $O(V^{-1})$ (see volume 1), and adding up $N$ of them results in a total shift $\propto N/V = O(1)$. For unitary $s$-wave interactions, with divergent scattering lengths, this is not so clear *a priori*, since energy shifts in the ground state within effective field theory give a single-particle energy shift of $O(V^{-2/3})$, and therefore a proper calculation is necessary to establish the order of the correction. Here, we study these shifts by deriving Fumi's theorem [1].

In the many-body problem, the ground state corresponds to a filled Fermi sea, with orbitals given by the interacting single-particle orbitals. If $M$ bound states exist, with bound state energies $E_1, E_2,...,E_M$, then the ground state energy is given by $\sum_{i=1}^{N} E_i$ if $N \leqslant M$, which is typically of $O(1)$ (for example, the one-dimensional $-1/|x|$ interaction), but not necessarily, if $1/M = 0$. The rest of the contribution to the ground state energy will come from scattering states. If the potential is spherically symmetric, then the boundary condition for the reduced radial wave functions is simply

$$u_{k\ell}(R_0) = 0, \tag{4.2}$$

which implies (see equation (3.29))

$$k = \frac{\pi p}{R_0} - \frac{\pi}{2R_0}\ell - \frac{\theta_\ell(k)}{R_0}, \tag{4.3}$$

where $p = 1, 2,....$ Just as we did for the virial expansion in chapter 3, we take advantage of the fact that $k \to \pi p / R_0$ as $R_0 \to \infty$, so we replace equation (4.3) with

$$k \approx \frac{\pi p}{R_0} - \frac{\pi}{2R_0}\ell - \frac{\theta_\ell(\pi p/R_0)}{R_0}. \tag{4.4}$$

The single-particle energies are given by

$$\varepsilon_{p,\ell} = \frac{\hbar^2 k^2}{2m}, \tag{4.5}$$

and the ground state energy $E_0$ becomes

$$E_0 = \frac{\hbar^2}{2m}\sum_p \sum_\ell (2\ell + 1)[k(p, \ell)^2]. \tag{4.6}$$

We are only interested in the energy shift in the thermodynamic limit, that is, $\delta E_0 = E_0 - (\hbar^2/2m)\sum_{p,\ell}(2\ell + 1)2(\pi p/R_0 - \pi\ell/(2R_0))\theta_\ell(\pi p/R_0)/R_0$, which becomes

$$\delta E_0 = -\frac{\hbar^2}{m\pi} \int_0^{k_F} dk\, k \sum_\ell (2\ell + 1)\theta_\ell(k). \tag{4.7}$$

If, for instance, interaction is predominantly *s*-wave, we may approximate the energy shift as

$$\delta E_0 \approx -\frac{\hbar^2}{m\pi} \int_0^{k_F} dk k \theta_0(k).$$

(4.8)

It is left as a problem to the reader to show that, in the unitary limit, the above relation remains valid and is of $O(1)$.

## 4.2 Fermionic polarons

We consider now a spin-polarized many-fermion system, say with spin-up. The fermions do not interact among themselves. On top of them, we consider a spin-down fermion, with the same mass, which interacts with each of the spin-up fermions via a pairwise potential. For simplicity, we shall assume that the interaction is low-energy effective field theory (LO-EFT) *s*-wave. We write down the bare Hamiltonian (with bare coupling constant $g_0$) of the system in the second quantization, which is the simplest way to treat the problem,

$$H = \sum_{\mathbf{k},\sigma}\varepsilon(\mathbf{k})c_{\mathbf{k}\sigma}^\dagger c_{\mathbf{k}\sigma} + \frac{g_0}{\mathcal{V}}\sum_{\mathbf{k}\mathbf{k}'\mathbf{q}}c_{\mathbf{k}+\mathbf{q}\uparrow}^\dagger c_{\mathbf{k}'-\mathbf{q}\downarrow}^\dagger c_{\mathbf{k}'\downarrow}c_{\mathbf{k}\uparrow}.$$

(4.9)

Above $c_{\mathbf{k}\sigma}$ annihilates a spin-$\sigma$ fermion, and $\varepsilon(\mathbf{k}) = \hbar^2k^2/2m$ is the kinetic energy dispersion.

We assume we have a many-body system of non-interacting spin-up fermions, and one spin-down fermion. Since for one impurity the statistics does not matter (it may as well be a boson but with the same mass as the fermions), we rename the operators as $c_{\mathbf{k}\uparrow} \equiv c_{\mathbf{k}}$ and $c_{\mathbf{k}\downarrow} \equiv a_{\mathbf{k}}$. To treat this problem, we use the variational method, although we must be careful because the Hamiltonian is *non-Hermitian* when acting on wave functions outside its Hilbert space, and every state that is not an exact eigenstate (or a combination thereof) is *not* in the Hilbert space. Recall that we saw this issue with the Hellmann–Feynman theorem in chapter 2 for this exact Hamiltonian. Therefore, it is *not* guaranteed that any variational calculation will give an upper bound to the ground state energy. What we can guarantee is that the ground state energy will be exact to whichever order in perturbation theory the trial wave function can reproduce. After this warning, we consider the following trial wave function, known as Chevy's ansatz [2]

$$|\psi\rangle = \left[\sqrt{Z_0} + \sum_{\mathbf{k}\notin F,\mathbf{q}\in F}\phi_{\mathbf{k},\mathbf{q}}c_{\mathbf{k}}^\dagger c_{\mathbf{q}}a_{\mathbf{q}-\mathbf{k}}^\dagger\right]|F\rangle,$$

(4.10)

where $|F\rangle \equiv |F\rangle_\uparrow \otimes a_0^\dagger|0\rangle_\downarrow$ is the non-interacting ground state. The right-expectation value of the Hamiltonian is given by

$$\langle H\rangle \equiv \langle\psi|H\psi\rangle \equiv \langle H_0\rangle + \langle V\rangle.$$

(4.11)

By working out the corresponding contractions, we obtain

$$\langle H_0 \rangle = \sum_{\mathbf{k} \notin F, \mathbf{q} \in F} \varepsilon_{\mathbf{kq}} |\phi_{\mathbf{kq}}|^2, \tag{4.12}$$

$$\langle V \rangle = \frac{g_0}{\mathcal{V}} \left[ \sum_{\mathbf{q} \in F} Z_0 + \sum_{\mathbf{k}, \mathbf{k}' \notin F, \mathbf{q} \in F} \phi^*_{\mathbf{k}, \mathbf{q}} \phi_{\mathbf{k}'\mathbf{q}} + \sum_{\mathbf{k} \notin F, \mathbf{q}, \mathbf{q}' \in F} \phi^*_{\mathbf{kq}'} \phi_{\mathbf{kq}} \right.$$
$$\left. + \sum_{\mathbf{k} \notin F, \mathbf{q} \in F} ((\sqrt{Z_0})^* \phi_{\mathbf{kq}} + \sqrt{Z_0} \phi^*_{\mathbf{kq}}) \right], \tag{4.13}$$

where $\varepsilon_{\mathbf{kq}} \equiv \varepsilon(\mathbf{k}) + \varepsilon(\mathbf{q} - \mathbf{k}) - \varepsilon(\mathbf{q})$. The functional we should minimize, fixing the normalization of the trial wave function, is given by

$$\mathcal{F}[\sqrt{Z_0}, (\sqrt{Z_0})^*, \phi, \phi^*] = \langle H \rangle - \mathcal{E}||\psi||^2, \tag{4.14}$$

where $\mathcal{E}$ will be the ground state energy and plays the role here of a Lagrange multiplier. Extremizing $\mathcal{F}$ with respect to $(\sqrt{Z_0})^*$ and $\phi^*$, the problem reduces to the following set of equations

$$\varepsilon_{\mathbf{kq}} \phi_{\mathbf{kq}} + \frac{g_0}{\mathcal{V}} \left[ \sqrt{Z_0} + \sum_{\mathbf{k}' \notin F} \phi_{\mathbf{k}'\mathbf{q}} \right] = E\phi_{\mathbf{kq}}, \tag{4.15}$$

$$\frac{g_0}{\mathcal{V}} \sum_{\mathbf{q} \in F} \left[ \sqrt{Z_0} + \sum_{\mathbf{k}' \notin F} \phi_{\mathbf{k}'\mathbf{q}} \right] = E\sqrt{Z_0}. \tag{4.16}$$

Above, we have defined $E = \mathcal{E} - \mathcal{E}_0$, where $\mathcal{E}_0$ is the ground state energy of the non-interacting system. Further defining $f(\mathbf{q}) = \sqrt{Z_0} + \sum_{\mathbf{k}' \notin F} \phi_{\mathbf{k}'\mathbf{q}}$, we find

$$f(\mathbf{q}) = \frac{\sqrt{Z_0}}{1 - \dfrac{g_0}{\mathcal{V}} \sum_{\mathbf{k} \notin F} \dfrac{1}{E - \varepsilon_{\mathbf{kq}}}}, \tag{4.17}$$

which is easy to eliminate in favour of a single non-linear equation for the ground state variational energy

$$E = \frac{1}{\mathcal{V}} \sum_{\mathbf{q} \in F} \left[ \frac{1}{g_0} - \frac{1}{\mathcal{V}} \sum_{\mathbf{k} \notin F} \frac{1}{E - \varepsilon_{\mathbf{kq}}} \right]^{-1} \rightarrow \int_{\mathbf{q} \in F} \frac{d\mathbf{q}}{(2\pi)^3} \left[ \frac{1}{g_0} - \int_{\mathbf{k} \notin F} \frac{d\mathbf{k}}{(2\pi)^3} \frac{1}{E - \varepsilon_{\mathbf{kq}}} \right]^{-1}, \tag{4.18}$$

In the last line above we have taken the infinite volume limit. Equation (4.18) can be easily solved numerically. Recall, however, that regularization–renormalization must be performed. The LO-EFT $s$-wave bare coupling $g_0$ as a function of the spherical cutoff $\Lambda$ is given by (see volume 1)

$$\frac{1}{g_0} = \frac{m}{4\pi\hbar^2 a} - \frac{m\Lambda}{2\pi^2\hbar^2}. \tag{4.19}$$

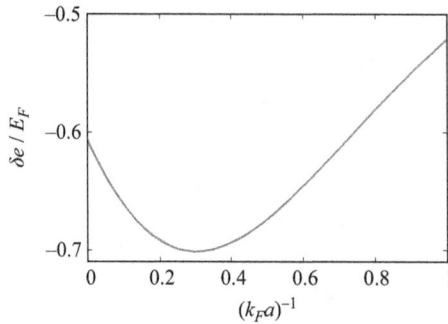

**Figure 4.1.** Energy with respect to the vacuum two-body energy $\delta E = (E + \hbar^2/ma^2)/E_F$ as a function of $1/k_F a$ for Chevy's polaron ansatz.

There is a similar associated problem when $k_F a \lesssim 1$, which is that of the molecule. Clearly, when $a \to 0^+$, there is a deeply bound state between the impurity and a fermion, so that the ground state energy approaches the bound-state energy in vacuum of the pair plus the energy of the Fermi sea with one less spin-up fermion,

$$E \to -\frac{\hbar^2 k_F^2}{2m} - \frac{\hbar^2}{ma^2}, \quad a \to 0^+. \tag{4.20}$$

The ground state energy for the polaron, from Chevy's trial wave function is plotted in figure 4.1, where we clearly see that the molecular energy must cross the polaron's energy and the polaron is not always the ground state.

## 4.3 Casimir interaction in mixtures of heavy and light fermions

We now consider a mixture of majority fermions with mass $m$ much lighter than the corresponding fermionic impurities, with mass $M$ ($M/m \gg 1$). In this case, instead of a single (polaron) impurity atom, we will deal with a multiple of them. We consider $s$-wave LO-EFT interactions between majority and minority fermions with scattering length $a$, and regard the light–light and heavy–heavy direct interactions as negligible. We follow reference [3].

Since the impurities are heavy, we will employ the Born–Oppenheimer approximation to establish their emergent effective interactions. We consider two heavy fermions and one light fermion with coordinates $r_1$, $r_2$ and $y$, respectively. The heavy fermions are fixed, and we calculate their energy shifts due to the presence of the light fermion, as a function of their interatomic distance. The Schrödinger equation for the light atom (in a scattering state) has the form

$$-\frac{\hbar^2 \nabla_y^2}{2m}\psi(y; r_1, r_2) = \frac{\hbar^2 k^2}{2m}\psi(y; r_1, r_2), \tag{4.21}$$

with the Bethe–Peierls boundary condition appropriate for LO interactions,

$$\psi(y; r_1, r_2) \propto \frac{1}{|r_i - y|} - \frac{1}{a}, \quad y \to r_i \ (i = 1, 2). \tag{4.22}$$

There are two solutions to the Schrödinger equation, with parities $P = \pm 1$, given by

$$\psi_\pm = \frac{\sin(k |\mathbf{r}_1 - \mathbf{y}| + \theta_\pm(\mathbf{r}_{12}))}{|\mathbf{r}_1 - \mathbf{y}|} \pm \frac{\sin(k |\mathbf{r}_2 - \mathbf{y}| + \theta_\pm(\mathbf{r}_{12}))}{|\mathbf{r}_2 - \mathbf{y}|}. \tag{4.23}$$

As $\mathbf{y} \rightarrow \mathbf{r}_1$, for instance, the above state becomes

$$\psi_\pm \propto \frac{1}{|\mathbf{r}_1 - \mathbf{y}|} \pm \frac{\sin(k r_{12} + \theta_\pm(\mathbf{r}_{12}))}{r_{12} \sin \theta_\pm(\mathbf{r}_{12})}. \tag{4.24}$$

Using the Bethe–Peierls boundary condition, equation (4.22), we find

$$\tan \theta_\pm(\mathbf{r}_{12}) = -\frac{k r_{12} \pm \sin(k r_{12})}{r_{12}/a \pm \cos(k r_{12})}. \tag{4.25}$$

To extract the energy shift—and therefore the effective interaction—for two impurities, we place the system in a sphere of radius $R$ with open boundary conditions at its surface. Without loss of generality, we assume $R \gg r_{12}$, place the impurities at $\mathbf{r}_1 = \mathbf{r}_{12}/2$, $\mathbf{r}_2 = -\mathbf{r}_{12}/2$, and apply the open boundary condition at a point $\mathbf{R}$ colinear with $\mathbf{r}_1$ and $\mathbf{r}_2$. Then, for positive parity states, we have

$$\sin(kR + \theta_+(r_{12})) = 0 \implies kR + \theta_+(r_{12}) = n\pi, \quad n = 1, 2, \ldots. \tag{4.26}$$

For negative parity states, the condition becomes instead

$$\cos(kR + \theta_-(r_{12})) = 0 \implies kR + \theta_-(r_{12}) = \left(n - \frac{1}{2}\right)\pi, \quad n = 1, 2, \ldots. \tag{4.27}$$

We have now everything in place to calculate the total energy shift for two impurities given a light fermion Fermi gas at finite density, with Fermi momentum $k_F$. We need to calculate the ground state energy in the presence of the two impurities, which we already did in section 4.1 with Fumi's theorem. In this case, the interatomic distance-dependent energy shift is given by (with obvious notation)

$$\begin{aligned}
E(r_{12}) &= \sum_{k_+} \frac{\hbar^2 k_+^2}{2m} + \sum_{k_-} \frac{\hbar^2 k_-^2}{2m} - E_0 \\
&= -\frac{\hbar^2}{m\pi} \int_0^{k_F} dk\, k\, [\theta_+(k, r_{12}) + \theta_-(k, r_{12})],
\end{aligned} \tag{4.28}$$

where $E_0$ is the non-interacting ground state energy of the majority Fermi gas, we have used Fumi's theorem, equation (4.8). Note that the LO-EFT interaction does support bound states (see problem 5), but we shall neglect them here for simplicity.

Let us study a simple example, namely the weakly interacting limit $a \rightarrow 0^-$. The phase shifts, equation (4.25), take the simple form

$$\tan \theta_\pm \approx -ka \pm \frac{\sin(k r_{12})a}{r_{12}}. \tag{4.29}$$

If $k_F a \to 0^-$, we may also substitute $\tan\theta_\pm \approx \theta_\pm$, so the energy shift, equation (4.28) is easily calculated as

$$E(r_{12}) \approx \frac{2}{3\pi} \frac{\hbar^2 k_F^2}{m} k_F a, \tag{4.30}$$

which is independent of distance. This must be the case, since a *single* heavy fermion does give an energy shift $\mu_h$, so that at large distances

$$E(r_{12}) \to 2\mu_h, \quad r_{12} \to \infty. \tag{4.31}$$

This means that the interaction potential must be shifted as

$$V(r_{12}) \equiv E(r_{12}) - 2\mu_h. \tag{4.32}$$

Now that this has been clarified, let us improve the calculation of the interaction potential in the limit $a \to 0^-$. In this limit, the phase shifts are expanded as

$$\theta_\pm(r_{12}) = -\frac{a}{r_{12}}\left[kr_{12} \pm \sin(kr_{12})\right] \pm \frac{a^2}{r_{12}^2}\left(kr_{12} \pm \sin(kr_{12})\right)\cos(kr_{12}) + O(a^3). \tag{4.33}$$

Introducing the above expression into Fumi's theorem, equation (4.28), and subtracting twice the chemical potential, we obtain

$$\frac{V(r_{12})}{(k_F a)^2} = \frac{\hbar^2 k_F^2}{2m} \frac{2k_F r_{12} \cos(2k_F r_{12}) - \sin(2k_F r_{12})}{2\pi(k_F r_{12})^4} + O(k_F a). \tag{4.34}$$

Note that the above interaction oscillates (it is of the RKKY type), and therefore the oscillations reduce the range of the effective power law tails from $r^{-3}$ to to $r^{-5}$ (see section 1.1). This means that the two-body scattering length (for heavy–heavy collisions) remains well defined. In figure 4.2, we plot the interaction in the limit $a \to 0^-$, where the oscillations and the power law tails are clearly visible.

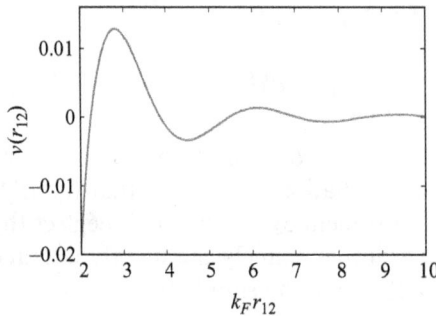

**Figure 4.2.** Effective interaction $v(r_{12}) = V(r_{12})/(E_F(k_F a)^2)$ between two heavy impurities in the limit $a \to 0^-$.

## Problems

1. Consider a non-interacting Bose gas in its ground state, interacting with a fixed impurity. Derive the equivalent of Fumi's theorem in this scenario, and apply it to the case of $s$-waves.

2. 
   (a) Consider a spinless Fermi gas in its ground state interacting with a fixed impurity. Assume the impurity–fermion interaction to be purely $s$-wave, and LO EFT, with infinite scattering length $1/a = 0^-$. Show that Fumi's theorem gives a finite $O(1)$ energy shift.

   (b) What happens if, instead of a Fermi gas, we consider a Bose gas?.

3. Consider a spin-polarized Fermi gas interacting via an $s$-wave LO EFT (scattering length $a$) with a mobile impurity with the same mass as the majority particles. Using Chevy's ansatz, obtain the expansion for the ground-state energy shift near the unitary limit. That is, if $E(0)$ is the shift for $1/a = 0$, obtain

$$E(a) = E(0) + \frac{c_1}{a} + \frac{c_2}{a} + O(a^{-3}). \tag{4.35}$$

4. Adapt Tan's relations to Fumi's theorem with LO EFT $s$-wave interactions, and show by an explicit calculation that the momentum distribution acquires $k^{-4}$ tails.

5. The LO EFT $s$-wave interaction may support bound states. Calculate the form of the bound-state contribution to the effective interaction between heavy polarons immersed in a Fermi sea of lighter majority particles.

6. Consider a chiral one-dimensional non-interacting Fermi gas, with linear dispersion $\varepsilon(k) = \hbar v k$. Assume it interacts with a fixed impurity via a potential of the form $V(x)$.
   (a) For $V(x)$ a regular interaction with $V(x) \to 0$ as $x \to \pm\infty$, obtain the equivalent of Fumi's theorem.
   (b) Do the same for $V(x) = g\delta(x)$, identifying any possible subtleties due to regularization and renormalization.

## References

[1] Mahan G D 2000 *Many-Particle Physics* (Dordrecht: Kluwer Academic)
[2] Chevy F 2006 Universal phase diagram of a strongly interacting Fermi gas with unbalanced spin populations *Phys. Rev.* A **74** 063628
[3] Nishida Y 2009 Casimir interaction among heavy fermions in the BCS-BEC crossover *Phys. Rev.* A **79** 013629

**IOP** Publishing

Strongly Interacting Quantum Systems, Volume 2
Many-body physics
**Manuel Valiente and Nikolaj Thomas Zinner**

# Chapter 5

## Strong microscopic interactions—weakly-coupled many-body systems

Throughout the book, we have encountered many examples for which, even though interactions appear strong (e.g. contain hard cores, or long-range tails), the system may behave as if it were weakly interacting or, more generally, we have been capable of establishing some non-perturbative effects without ever needing to solve a strongly interacting many-body system, and only working out the few-body problem, which is far more manageable. In this chapter, we put some of these fortunate issues into context, and study a few simple examples for which many-body physics with strong microscopic interactions can be quantitatively and qualitatively described by weakly coupled theories. We begin with a very well-known lattice model with strong interactions, and show that in some instances it can be treated as dilute continuum Bose gas. We then study a familiar one-dimensional example directly in the continuum.

### 5.1 Three-dimensional spin model in the dilute limit

We consider a cubic lattice in three spatial dimensions, with lattice spacing $d$. On each point of the lattice, we place a spin, either $\uparrow$ or $\downarrow$ (spin-1/2). The coupling between spins on the lattice only occurs between nearest neighbours. The Hamiltonian is given by

$$H = -J\sum_{\langle i,j\rangle}\mathbf{S}_i \cdot \mathbf{S}_j + V\sum_{\langle i,j\rangle}S_i^z S_j^z + \lambda\sum_i S_i^z. \tag{5.1}$$

Above, $\mathbf{i}$ and $\mathbf{j}$ are in $\mathbb{Z}^3$, $J, V$ and $\lambda$ are parameters, and $\mathbf{S}_i = (S_i^x, S_i^y, S_i^z)$ are spin-1/2 operators. Hamiltonian (5.1) is in general very difficult to handle analytically. Before looking at cases in which this is possible, let us develop equation (5.1) by writing it in terms of the Pauli matrices $\vec{\sigma_i}$ and, more conveniently, in terms of the ladder operators $\sigma_i^{\pm}$ and the $z$-component of the spin $\sigma_i^z$. We define

doi:10.1088/978-0-7503-3091-6ch5

$$t \equiv \frac{J\hbar^2}{8},$$ (5.2)

$$v \equiv \frac{\hbar^2 V}{4} - 2t,$$ (5.3)

$$\gamma \equiv \frac{\hbar\lambda}{2},$$ (5.4)

and the Hamiltonian becomes

$$H = -t\sum_{\langle i,j \rangle}\left[\sigma_i^+\sigma_j^- + \sigma_j^+\sigma_i^-\right] + v\sum_{\langle i,j \rangle}\sigma_i^z\sigma_j^z + \gamma\sum_i\sigma_i^z.$$ (5.5)

Since $\sigma^z = \sum_i\sigma_i^z$ is conserved, we work in subspaces with fixed magnetization. As a reference state, we consider the vacuum $|0\rangle \equiv |\ ...\ \downarrow\downarrow\ ...\ \downarrow\ ...\ \rangle$, that is, the minimal spin projection (totally ferromagnetic state). It is an eigenstate of the Hamiltonian, with eigenenergy $E_0$ given by (as the number of sites $L_s^3 \to \infty$)

$$\frac{E_0}{L_s^3} = 6v - \gamma.$$ (5.6)

We shall measure every energy with $E_0$ as a reference. The simplest case we can consider consists of placing $v \equiv 0$, and $\gamma = +6t$. This ensures that the single-magnon ground state energy vanishes. The single-magnon ground state is simply

$$|\psi_1\rangle = \sum_i\sigma_i^+|0\rangle \equiv \sum_i|\uparrow\rangle_i,$$ (5.7)

and its energy is zero with respect to the vacuum energy $E_0$. We now consider the two-magnon problem. Since the spin Hamiltonian is equivalent to hard-core bosons (the hard-core bosons are the magnons), we can simply use the two-body solution for on-site interactions $U$ with $U \to \infty$ (see volume 1). The $T$-matrix is a constant (quasi-momentum independent), and takes the form, for total momentum $\mathbf{K} \equiv 0$,

$$\frac{1}{T(z)} = \frac{1}{Ud^3} - \int_{BZ}\frac{d^3k}{(2\pi d)^3}\frac{1}{z - 2\varepsilon(\mathbf{k})},$$ (5.8)

where $\varepsilon(\mathbf{k}) = -2t\sum_{l=x,y,z}(\cos(k_l d) - 1)$ is the three-dimensional dispersion. In our case, the hardcore conditon further simplifies equation (5.8), since $1/U = 0$. We will be interested in a dilute magnon gas, so that the main interaction parameter is the scattering length, and we set $z = 0$ in equation (5.8). Fortunately, the remaining integral is a Watson's integral, with value

$$\frac{W}{4t} = \int_{BZ}\frac{d^3k}{(2\pi)^3}\frac{1}{2\varepsilon(\mathbf{k})},$$ (5.9)

and $W = 0.505462....$ Therefore, the $T$-matrix at zero energy and in the hard core limit is given by

$$\frac{1}{T(0)} = \frac{0.1263655\ldots}{td^3}. \tag{5.10}$$

Near the continuum limit, $t = \hbar^2/2md^2$, with $m$ the effective mass, for a non-relativistic system and, since at zero energy the $T$-matrix is simply $T(0) = 4\pi\hbar^2 a/m$, we have

$$a/d = 0.31487\ldots . \tag{5.11}$$

Therefore, for small but non-vanishing lattice spacings, and low magnon density, these are described by a repulsive non-relativistic Bose gas with scattering length $a = \alpha d$, with $\alpha \approx 0.315$. This allows us to use Bogoliubov theory in the dilute limit and, if $N$ is the number of magnons, we identify the dilute limit as

$$\rho a^3 \ll 1 \implies \frac{N}{L_s^3}\alpha^3 \ll 1 \implies \frac{N}{L_s^3} \ll 32. \tag{5.12}$$

which is *independent* of the lattice spacing. Note that the above condition is always satisfied since the number of magnons per site is strictly smaller than 1. We need to supplement this condition with a condition on the appropriateness of the non-relativistic approximation. To be conservative, we may use free fermions to estimate the maximal magnonic filling factor. The highest momentum state occupied by a fermion would have $|\mathbf{k}| = k_F d \sim 2\pi N/L_s$. Its energy on the lattice would be $\sim -2t(\cos(k_F d) - 1)$. If we set the maximal deviation from the non-relativistic dispersion to 1%, then we need $|1 + 2(\cos(2\pi N/L_s) - 1)/(2\pi N/L_s)^2| < 0.01$, which gives $N/L_s^3 < 0.06$. This is a very safe estimate, but it gives a rough idea of the validity of the non-relativistic and dilute approximations altogether. For moderate particle numbers, the spin model can be simulated using quantum Monte Carlo methods. With, say, 100 particles and right at the limit $N/L_s^3 = 0.06$, a cubic lattice with side length $L_s \approx 12$ is doable.

This approach shows that, while a system may be strongly interacting, a weak-coupling theory can be used to describe it. In this case, the Bogoliubov approximation comes in handy. For simplicity, let us simply quote the Bogoliubov Hamiltonian, which takes the form [1]

$$H_B = \sum_{k \neq 0}\frac{\hbar^2 \mathbf{k}^2}{2m}a_\mathbf{k}^\dagger a_\mathbf{k} + \frac{g_0\rho}{2}$$
$$\sum_{k \neq 0}\left[(a_\mathbf{k}^\dagger a_\mathbf{k} + a_{-\mathbf{k}}^\dagger a_{-\mathbf{k}}) + (a_\mathbf{k}^\dagger a_{-\mathbf{k}}^\dagger + a_\mathbf{k}a_{-\mathbf{k}})\right] + \frac{1}{2}Vg_0\rho^2, \tag{5.13}$$

where $a_\mathbf{k}^\dagger$ and $a_\mathbf{k}$ are bosonic creation and annihilation operators. Since equation (5.13) is a canonical quadratic form, it is diagonalized by using a Bogoliubov transformation. Here, we only wish to discuss the ground state energy and the renormalization of the interaction. The energy per particle in the ground state is given by

$$\frac{E}{N} = \frac{g_0 \rho}{2} - \frac{1}{2\rho} \int \frac{d\mathbf{k}}{(2\pi)^3} \left[ \frac{\hbar^2 \mathbf{k}^2}{2m} + g_0 \rho - \mathcal{E}(\mathbf{k}) \right], \tag{5.14}$$

where $\mathcal{E}(\mathbf{k}) = \sqrt{g_0 \rho \hbar^2 k^2/m + (\hbar^2 k^2/2m)^2}$ is the quasi-particle energy. We now need to show that equation (5.14) is finite, within a consistent renormalization scheme. Since Bogoliubov theory is a second-order perturbation theory for Bose gases, we should renormalize the bare coupling constant to second order. This is done in detail in volume 1. The bare coupling constant is expanded as

$$g_0 = g_R - g_R^2 \int \frac{d\mathbf{k}}{(2\pi)^3} \frac{1}{\hbar^2 k^2/m} + O(g_R^3) \equiv g_R + C_2 g_R^2 + O(g_R^3), \tag{5.15}$$

where $g_R = 4\pi\hbar^2 a/m$ is the renormalized coupling constant for LO-EFT $s$-wave interactions. We now expand the quasi-particle energy to second order in the bare coupling constant, obtaining

$$\begin{aligned} \mathcal{E}(\mathbf{k}) &= \frac{\hbar^2 k^2}{2m} + g_0 \rho - \frac{g_0^2 \rho^2}{\hbar^2 k^2/m} + O(g_0^3) \\ &= \frac{\hbar^2 k^2}{2m} + g_R \rho + C_2 \rho g_R^2 - \frac{g_R^2 \rho^2}{\hbar^2 k^2/m} + O(g_R^3). \end{aligned} \tag{5.16}$$

Inserting these expressions into the ground state energy per particle, equation (5.14), we obtain

$$\frac{E}{N} = \frac{g_R \rho}{2} + C_2 \frac{g_R^2}{2} - \frac{1}{2\rho} \int \frac{d\mathbf{k}}{(2\pi)^3} \left[ \frac{g_R^2 \rho^2}{\hbar^2 k^2/m} \right] + O(g_R^3) = \frac{g_R \rho}{2} + O(g_R^3). \tag{5.17}$$

That is, the usual renormalization procedure makes the Bogoliubov ground state energy per particle finite. This is what we wanted to show, and we leave the details of higher order corrections to the interested reader (see [1]).

## 5.2 Resonant one-dimensional bosons

We study now a very different example, where interactions are typically strong, but the system is effectively non-interacting. Some residual three-body interactions will remain, but these are typically feeble and can be understood perturbatively. The particular system of interest is that of hard-core (or almost hard-core) bosons in one spatial dimension. If these are hard core, then they are equivalent to spinless fermions and the results below apply to them as well. Their Hamiltonian is given by

$$H = \sum_{i=1}^{N} \frac{p_i^2}{2m} + \sum_{i<j=1}^{N} V(x_i - x_j). \tag{5.18}$$

We place the system at a two-boson zero-energy resonance, that is, the even-wave scattering length diverges $(1/a = 0)$. Then, asymptotically, the two-boson zero-energy state takes the form, in the relative coordinate,

$$\psi(x) \propto 1, \quad |x| \to \infty. \tag{5.19}$$

At higher energies, the effective range expansion for the even-wave phase shifts is given by (see volume 1)

$$-k \cot \theta(k) = \frac{1}{a} - \frac{1}{2}rk^2 + O(k^4). \tag{5.20}$$

On resonance, $\theta(k) = \pi/2$, obviously. For $k \neq 0$, we have, to leading order, $k \cot \theta(k) = rk^2/2$. To see how weak the effect of effective range is, consider the two-boson lowest-energy state in a ring of length $L$ with vanishing total momentum $K$. From equation (2.222), we obtain

$$k = \frac{\pi}{L} + \frac{\pi r}{L^2} + O(L^{-3}). \tag{5.21}$$

That is, the correction due to range is one order lower than the leading correction, since its energy is

$$\frac{\hbar^2 k^2}{m} = \frac{\pi^2}{L^2} + \frac{2\pi^2 r}{L^3} + O(L^{-4}). \tag{5.22}$$

The effective range correction turns out to be of lower order than the correction due to the emergent three-body interaction. In the weak-coupling limit, the energy shift is given by (see volume 1)

$$E = \frac{\hbar^2}{2mL^2} \frac{3\pi^2}{\log|Q_* L|} + O(L^2 \log(|Q_* L|)^{-2}), \tag{5.23}$$

where $Q_*$ is the three-body parameter. While the power $L^{-2}$ is suppressed due to a quantum anomaly (renormalization of a scale invariant interaction breaking the scale invariance), the suppression is logarithmic, which is much slower than the linear suppression for effective range. Therefore, we cannot neglect the three-body interaction in a resonant one-dimensional gas, and it has to be included at a lower order than the effective range. In fact, it is *necessary* in order for the ground state energy not to vanish in a many-body system and for effective range corrections to be of any effect. To finalize this section, we perform a mean-field calculation for resonant bosons. Since the interaction is kinematically equivalent to a two-dimensional delta interaction, the mean-field correction to the ground state energy (which vanishes otherwise) requires some care. First, we define the three-body scattering length $a_3$ as

$$(Q_* a_3)^2 = 8e^{-2\gamma}, \tag{5.24}$$

where $\gamma$ is Euler's constant. The energy shift, including the finite density, is therefore

$$E \propto \frac{\hbar^2 \rho^2}{m} \frac{1}{\log(\rho a_3)}, \tag{5.25}$$

where the proportionality constant is ambiguous due to the renormalization procedure. This is simply a matter of scale, and is in principle not a problem if higher order terms are included. Of course, our correction in equation (5.25) coincides with the leading order in the full many-body theory [2].

## Problems

1. Consider the spin model in equation (5.5), with $\gamma = 6t$. Allow $v$ to be non-zero, and identify the weak-coupling limits of the theory as a function of $v$. (Hint: There may be several weak-coupling limits.)
2. Consider three identical particles in one spatial dimension interacting via a hard-core three-body interaction of the form

$$V(x_1, x_2, x_3) = V_0 \Theta(a_3 - \rho),$$

where $V_0 \to \infty$, $a_3 > 0$ and $\rho^2 = x_{12}^2 + x_{13}^2 + x_{23}^2$.
   (a) Solve the three-boson scattering problem and identify the relation between $a_3$ and the three-body scattering length.
   (b) Solve the three-fermion problem. Is there a difference between bosons and fermions? If so, why?
   (c) Go beyond the second-order virial expansion to obtain the leading order correction to the pressure for both the bosonic and fermionic systems.

## References

[1] Fetter A and Walecka J D 2003 *Quantum Theory of Many-Particle Systems* (New York: Dover)
[2] Pastukhov V 2019 Ground-state properties of dilute one-dimensional Bose gas with three-body repulsion *Phys. Lett.* **383** 894

**IOP** Publishing

Strongly Interacting Quantum Systems, Volume 2
Many-body physics
**Manuel Valiente and Nikolaj Thomas Zinner**

# Appendix A

## Definitions of correlation functions

In this appendix, we set the conventions for different correlation functions used throughout the book

### A.1 $M$-body distribution function

For a system of $N$ identical, spinless particles, we define the $M$-body ($M < N$) distribution function as the marginal probability density function evaluated at positions $r'_1,...,r'_M$. If the wave function in the first quantization and in the position representation is $\psi(r_1,...,r_N)$, then the $M$-body distribution function is defined as

$$g_M(r'_1,...,r'_M) = \frac{N!}{(N-M)!} \int dr_{M+1} \ldots dr_N |\psi(r'_1,...r'_M, r_{M+1},...,r_N)|^2. \quad (A.1)$$

In operator form, this is simply given by $g_M = \langle \hat{g}_M(s_1,...,r_M) \rangle$, with

$$\hat{g}_M(r_1,...,r_M) = \sum_{i_1,...,i_M}' \delta(r_{i_1} - r'_1)\delta(r_{i_2} - r'_2) \ldots \delta(r_{i_M} - r'_M), \quad (A.2)$$

where the primed sum indicates that $i_j \neq i_l$ for $j \neq l$.

We consider now a system of $N$ spin-1/2 fermions, with $N_\uparrow$ fermions with $\sigma_z = +1$ and $N_\downarrow = N - N_\uparrow$ fermions with $\sigma_z = -1$. The wave function is given by $\psi(r_1\sigma_1,...,...,r_N\sigma_N)$. In the book, we only use the pair distribution function, which we call $g(r, r') \equiv g_2(r, r')$. It is given by

$$g(r, r') = \langle \hat{g}_2(r, r') \rangle = g_{\uparrow\uparrow}(r, r') + g_{\downarrow\downarrow}(r, r') + g_{\uparrow\downarrow}(r, r'), \quad (A.3)$$

where

$$g_{\sigma\sigma}(r, r') \equiv \frac{N_\sigma!}{(N_\sigma - 2)!} \int dr_3,...,dr_N |\psi(r\sigma, r'\sigma, r_3\sigma_3,...,r_N\sigma_N)|^2, \quad (A.4)$$

doi:10.1088/978-0-7503-3091-6ch6      A-1      © IOP Publishing Ltd 2025. All rights,

$$g_{\uparrow\downarrow}(\mathbf{r}, \mathbf{r}') \equiv N_\uparrow N_\downarrow \int d\mathbf{r}_3,\ldots,d\mathbf{r}_N |\psi(\mathbf{r}\uparrow, \mathbf{r}'\downarrow, \mathbf{r}_3\sigma_3,\ldots,\mathbf{r}_N\sigma_N)|^2. \tag{A.5}$$

For homogeneous systems, it is convenient to define

$$g_2(\mathbf{r}) = \int d\mathbf{R} g_2(\mathbf{R} - \mathbf{r}/2, \mathbf{R} + \mathbf{r}/2), \tag{A.6}$$

$$g_{\sigma\sigma'}(\mathbf{r}) = \int d\mathbf{R} g_{\sigma\sigma'}(\mathbf{R} - \mathbf{r}/2, \mathbf{R} + \mathbf{r}/2), \tag{A.7}$$

where a slight abuse of notation seems harmless given the number of arguments in each function.

On a one-dimensional lattice, for spinless bosons or fermions, in which the positions the particles can occupy are $x_i \in \mathbb{Z}d$, with $d$ the lattice spacing, the two-body correlation function is defined as

$$g_2(x, x') = N(N - 1)d^{N-2} \sum_{x_3,\ldots,x_N} |\psi(x, x', x_3,\ldots,x_N)|^2. \tag{A.8}$$

Here, the normalization condition for the wave function is

$$d^N \sum_{x_1,\ldots,x_N} |\psi(x_1,\ldots,x_N)|^2 = 1, \tag{A.9}$$

and one-dimensional, one-particle scalar products are defined via

$$\langle \phi|\psi \rangle = d\sum_x \phi^*(x)\psi(x). \tag{A.10}$$

This notation emphasizes the pass to the continuum limit ($d \to 0$), where

$$\langle \phi|\psi \rangle \to \int dx \phi^*(x)\psi(x), \quad d \to 0. \tag{A.11}$$

## A.2 Pair density function

For completeness, we include the definition of the pair density function, and relate it to the pair distribution function when the comparison is relevant, as well as to examine notation regarding short-range correlations and the contact. The pair density function is defined via $\rho_2^{(N)}(r) = \langle \hat{\rho}_2^{(N)}(r) \rangle$, with

$$\hat{\rho}_2^{(N)}(r) = \frac{1}{r^2} \sum_{i<j=1}^N \delta(r_{ij} - r). \tag{A.12}$$

For a homogeneous, spherically symmetric system of identical particles, we obtain

$$\rho_2^{(N)}(r) = 2\pi g_2(r). \tag{A.13}$$

The pair density for a two-body system with relative wave function $\phi_2(\mathbf{r})$, which we denote $\rho_2(r)$, is defined as

$$\rho_2(r) = \int d\Omega |\phi_2(\mathbf{r})|^2 = |R(r)|^2, \qquad (A.14)$$

with the last equality being valid for solid-angle normalized spherically symmetric systems.

For bosons, in chapter 1 we found that $\rho_2^{(N)}(r) \to C_2^{(N)} \rho_2(r)$ which, for the pair distribution function $g_2(r)$ implies

$$g_2(r) \to C_2^{(N)} \frac{|R(r)|^2}{2\pi}. \qquad (A.15)$$

## A.3 Static structure factor

For a spinless system, we define the static structure factor as

$$S(\mathbf{k}) = 1 + \frac{1}{N} \int d\mathbf{r} d\mathbf{r}' e^{-i\mathbf{k}\cdot(\mathbf{r}-\mathbf{r}')} [g_2(\mathbf{r}, \mathbf{r}') - g_1(\mathbf{r})g_1(\mathbf{r}')]. \qquad (A.16)$$

If the system is homogeneous, then the density $g_1(\mathbf{r}) \equiv \rho$ is constant, and changing integration variables to centre of mass $(\mathbf{R} = (\mathbf{r} + \mathbf{r}')/2)$ and relative $(\mathbf{s} = \mathbf{r} - \mathbf{r}')$ coordinates, the structure factor becomes

$$S(\mathbf{k}) = 1 + \frac{1}{N} \int d\mathbf{s} e^{-i\mathbf{k}\cdot\mathbf{s}} \int d\mathbf{R} [g_2(\mathbf{R} - \mathbf{s}/2, \mathbf{R} + \mathbf{s}/2) - \rho^2]. \qquad (A.17)$$

Since $g_2(\mathbf{r}, \mathbf{r}') \to \rho^2$ as $\mathbf{s} \to \infty$, we define $\tilde{g}_2(\mathbf{r}) \equiv g_2(\mathbf{r}) - \rho^2$, and we obtain

$$S(\mathbf{k}) = 1 + \frac{1}{N} \int d\mathbf{r} e^{-i\mathbf{k}\cdot\mathbf{r}} \tilde{g}_2(\mathbf{r}), \qquad (A.18)$$

valid for a homogeneous system. If, moreover, the interaction between particles is spherically symmetric, so is $g_2(\mathbf{r}) \equiv g_2(r)$, and the structure factor is simplified to

$$S(k) = 1 + \frac{4\pi}{Nk} \int dr\, r \sin(kr) \tilde{g}_2(r). \qquad (A.19)$$

www.ingramcontent.com/pod-product-compliance
Lightning Source LLC
Chambersburg PA
CBHW082106210326
41599CB00033B/6600